W9-AHG-876

Yes, We Have No Neutrons

Yes, We Have No Neutrons

An Eye-Opening Tour through the Twists and Turns of Bad Science

A. K. Dewdney

John Wiley & Sons, Inc.

New York • Chichester • Weinheim • Brisbane • Singapore • Toronto

This text is printed on acid-free paper.

Published by John Wiley & Sons, Inc.

Library of Congress Cataloging-in-Publication Data

Dewdney, A. K.
Yes, we have no neutrons : an eye-opening tour through the twists and turns of bad science / A. K. Dewdney.
p. cm.
Includes index.
ISBN 0-471-10806-5 (cloth : alk. paper)
1. Errors, Scientific—Popular works. I. Title.
Q172.5.E77B48 1997
500—dc20 96-35312

Printed in the United States of America

10 9 8 7 6 5 4 3 2 1

Dedicated to Martin Gardner,
who actually believes in an
independent reality.

Contents

YES, WE HAVE NO NEUTRONS

Introduction
Of Neutrons, Sorcerers, and Apprentices

On March 23, 1989, the world was stunned to learn that two scientists named Fleishmann and Pons had achieved nuclear fusion in an apparatus hardly more complicated than a jar with some wires in it. It was "cold fusion," as the media explained. No huge billion-dollar "hot" reactors would be necessary. (They didn't work anyway.) The world stood on the threshold of a new age of incredibly cheap energy. For a few nights and days we lived in an atmosphere of dreamy unreality. Imagine! Free energy! It was magic.

Science is magic. The unvarying bend of the electron beam in a magnetic field, the hydrogen that never fails to bubble from the electrolyte, all the repeatable phenomena of this universe point to underlying laws that science seeks to know.

But magic can go awry, as the sorcerer's apprentice discovered. Failure to follow the exacting methodology of good science can produce strange results, as Fleishmann and Pons also discovered. At first they declared that their miraculous process was producing neutrons. Later, it appeared that neutrons were not exactly pouring forth from their humble apparatus, prompting the title of this book.

Yes, We Have No Neutrons does not address fraudulent science, but mistaken science. The appropriate persona, therefore, is not the con artist, but the bumbler, the apprentice if you will.

In learning the laws that govern physical reality and predicting new phenomena from them, the scientist seems like a sorcerer. Science, meanwhile, spawns technology that brings to life the tales of old; flying through the air, speaking over thousands of miles, the power of healing. These modern manifestations of science would have struck the ancients as sorcery of a high order. In short, they would have understood today's science about as well as today's general public.

When science goes wrong, all hell breaks loose. Claims are met with counterclaims and the public grows confused. Weren't we promised free energy? Those who set such debacles in motion pay dearly for their mistakes. Their reputations suffer and science suffers, too. The public grows skeptical about the scientific process and those who would paint science as a purely social activity with no meaningful truths behind it giggle with delight.

Apprentices and Sorcerers

Goethe wrote about an apprentice who copied a sorcerer's spell but didn't quite get it right. Using the spell, he charged his broom to replace him in the drudgeries of apprenticeship, including the task of carrying water. Things got out of hand when the broom refused to quit filling the watering tank after it was full. The apprentice resorted, finally, to an axe. But each splinter of the old broom became a new one, sprouting arms, legs, and a new pail to carry water.

Walt Disney brought Goethe's account to life in the animated film classic, *Fantasia*. Mickey Mouse played the apprentice. Who does not remember the old castle, Mickey clad in the sorcerer's robe and hat, the psychedelic armies of brooms, the relentless march of the Dukas symphony? Only when the castle was aflood did the sorcerer wake up and dry it with a spell. Mickey got off lightly with a swat of the broom.

A last word about Mickey. After setting the broom to work, he fell asleep to dream of controlling the universe. Dreams of glory will play their role in the chapters to come; so will the media, ironically enough. After fueling the thirst for fame, newspapers and television advertise apprentices' "discoveries" with embarrassing regularity. Perhaps they don't realize how difficult it can sometimes be to distinguish good science from bad.

If science is sorcery (of sorts) and scientists are sorcerers, who are the apprentices? Sometimes they are amateurs, eager to look like scientists, or people with some scientific training working outside their areas of expertise.

The people responsible for Biosphere 2, sometimes referred to as the disaster in the desert, appear to have been ill-prepared scientifically for the task they set for themselves. Their keen interest in scientific trappings, like red astronaut suits and the magnificent steel and glass structure that housed seven "bionauts" for a year in Arizona may have caught the media off guard. Television, in particular, is always on the lookout for stuff that looks scientific.

Sometimes the apprentices are genuine scientists whose research goes awry when they neglect a key ingredient of the scientific method. The scientists who announced cold fusion to a stunned world were respected chemists. But in their eagerness to be first to announce this startling new "discovery," they ran their research program off the rails, ignoring the fact that their experimental results were essentially irreproducible. A more careful methodology would have raised serious doubts that they had, indeed, achieved nuclear fusion. This book has two other examples of respected scientists behaving like apprentices: At the turn of the century, French physicist René Blondlot found rays that didn't exist; more recently, a handful of dedicated radio astronomers have been searching for signals from aliens. The aliens may or may not be there, but as it stands, the hypothesis (Aliens are sending us radio signals) is nonfalsifiable.

Between the bumbling of apprentices and the astonishing lapses of real sorcerers is a grey area that can be analyzed in two different ways. Were the psychologists who introduced IQ tests as a measure of "intelligence" sorcerers-gone-astray or were they essentially apprentices sending an army of brooms marching into the public psyche?

The physicist Richard Feynmann once described psychology and its relatives as "cargo cult science." He was referring to a tribe of native New Guineans who allegedly built crude wooden airplanes to attract gifts from the gods of the sky. During World War II, after all, the natives had seen tons of goods arrive at military airbases. Will the appearance of numerous mathematical formulae and scientific-sounding jargon in a paper bring real results from the sky? Perhaps the runways of social science are merely lined with its own cargo—crates of scientific papers.

This characterization goes too far for the social sciences as a whole, but it certainly captures many infamous avenues of inquiry, including the intelligence quotient. These were apprentices from the start, it would seem. And bad science begat more bad science when IQ became the basis for peculiar racial distinctions, as demonstrated by the authors of *The Bell Curve*.

Whether you look at Biosphere 2, cold fusion, or the theory of racial differences in intelligence, you're witnessing a major blow to the credibility of science, a wash of media interest that floods the old castle with unwelcome attention. And all too often the real sorcerers stay fast asleep.

It is time to explore the castle of science. I have three rusty old keys for

THE THREE DOORS

To understand what science is and how it works, we must pass through three doors of understanding. Behind each door is a major distinction that will clarify the chapters to follow. The first door is clearly labeled:

Science and Technology

When we open this door we find a vast hall of artifacts and machines:

> antennas, bicycles, computers, dental floss, electric drills, face masks, generators, hangers, igloos, jackhammers, kettles, lawn mowers, modems, nuclear reactors, oboes, power shovels, quake detectors, refrigerators, satellites, telephones, urinals, voting machines, wet suits, X-ray machines, Yahtzee games, and xylophones,

to name a few. Throughout the hall a monotonous voice drones on and on:

"Technology is not science and science is not technology." As indeed, it isn't, or, they aren't. The distinction sounds trite until you realize how confused we have all become by television programs that make no distinction at all. We are shown a wonderful one-off piece of technology like a Tokamak reactor, only to be told, This is science. Well, this is certainly a suitable symbol of science, and big science at that, but it is no more science than a frying pan is a full stomach.

To be a bit more exact, technology consists of objects that have specific purposes. Science consists of methods and results.

The confusion about what science is has spread to the universities, of all places. For the last few decades, university departments have been busily changing their names by adding the word "science." Engineering science, for example, is not a actually a science, but the study of technology and its techniques. Now there is management science, secretarial science, and even political science. The field we used to call physical education wanted to call itself physical science, but the name was already taken, so it settled for "kinesiology." Everybody wants to be a scientist.

The next door bears the austere title

The Two Sciences

Here, in a dank candlelit chamber, are Euclid, the Greek philosopher-mathematician, and Francis Bacon, the seventeenth-century English savant. Euclid stands before a blackboard on which he has drawn a triangle. At the moment he is laboriously writing out a proof that the sum of the interior angles of the triangle equals two right-angles, or 180 degrees. Bacon, meanwhile, sits before a bell jar, inside of which a candle grows more feeble by the moment, finally dying out in the exhausted atmosphere. Bacon makes a note on a piece of paper, removes the bell jar, ignites the candle anew with a cigarette lighter, and replaces the bell jar. He stares intently at the candle flame, which once more slowly dies away.

The two scientists exemplify deductive and inductive science, respectively. The deductive sciences include mathematics, applied mathematics, and the mathematical areas of computer science. Deductive scientists put their ideas to the test by trying to prove them deductively. A valid proof, like the one Euclid is trying to construct, consists of a series of steps, each derived from a previous step by the application of logic.

I ask Euclid, in flawless classical Greek (writing books gives you strange powers) to rephrase his proof in English. He concurs, but asks me to explain that the diagram represents an arbitrary triangle that lies in a perfectly flat plane. The base of the triangle has been extended beyond the two corners. Euclid has added a line parallel to the base and passing through the third corner. He has, in addition, labeled angles in the figure with the letters *a*, *b*, *c*, and *d*.

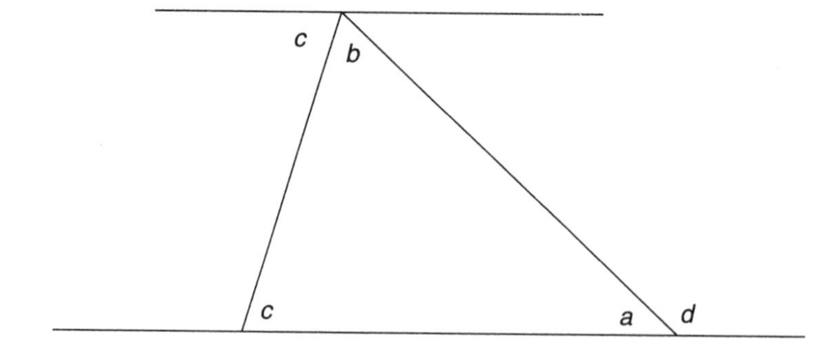

He writes out the following proof:

> The angles a and d sum to two right angles (180 degrees).
>
> The two horizontal lines are parallel, so by a previous theorem, the two angles marked c are equal.
>
> By the same theorem, the angle $b + c$ must equal the angle d.
>
> But if $a + d$ equals two right angles and d equals $b + c$, then $a + b + c$ must equal two right angles.

Euclid beams, and tears come to his eyes. "It is so beautiful," he says. "To think that eternal truths come purely by taking thought seems a gift from the gods."

All deductive science shares this character. Truths are established by the stepwise application of logic, like a game of chess. If the steps are all correct, the conclusion is certain. Some of the steps, like the two middle ones in Euclid's proof, appeal to previous results that are also obtained by deduction. Thus is the vast edifice of deductive science constructed.

When we turn to Francis Bacon, I ask the obvious question in seventeenth-century English, "When finish ye this experimente?"

"I know notte," replies Bacon, "for I know notte whyne but the candelle shalle not extinguishe itselfe, but continue in the burnynge."

The other major branch of science is the inductive branch, which includes sciences such as physics and biology. Here scientists try to prove their hypotheses by induction. If Bacon observes the candle die out inside the bell jar, he may feel compelled to repeat the experiment. If the flame dies a second time, then a third, he feels increasingly convinced that the candle will always flicker out under these circumstances. He might try the experiment a hundred times, confirming that the experiment is repeatable, always with the same result. This will help to convince others that a general law is involved. But is it just possible that on the hundred and first trial, the candle will go on merrily burning, even though all conditions were exactly what they were before? This is the burning question for Bacon, a purist. Most other inductive scientists would have stopped by now.

A deductive scientist has no such questions to fear. The sum of two positive integers is always positive, for example. You could, of course, experiment by trying all sorts of positive integers to see whether their sums turned out to be positive. But why use induction when you can use deduction? A mathematician can prove that the sum of two positive integers will always be a positive integer.

The tiniest possibility that observations might not always repeat themselves gives inductive science a somewhat exciting and teetery quality. Not all phenomena are equally predictable.

The inductive sciences have a rank structure, of sorts. It has been common practice to arrange these sciences on a spectrum, from "hard" to "soft."

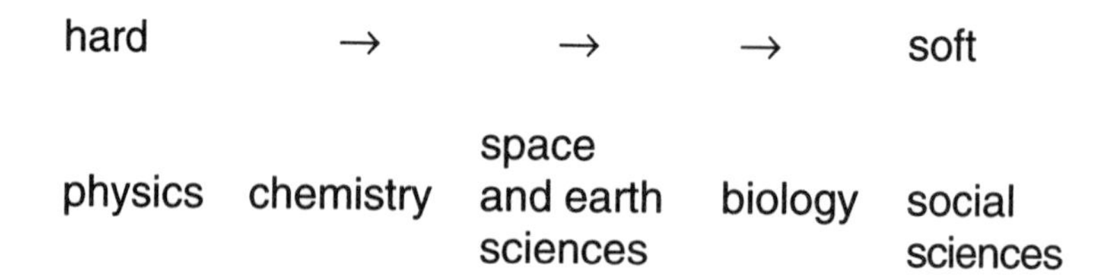

The words hard and soft refer not to the relative difficulty of the sciences in this spectrum, but to the relative certainty of their conclusions. The word hard has much the same meaning as the popular expression a "hard result" or "hard evidence." Although soft results may contain more than a grain of truth, they must sometimes be taken with a grain of salt.

Examples of bad deductive science are very rare, and there is a reason for this. The only place a deductive scientist can go badly wrong is in the proof. Any proof clear enough to publish is also clear enough to find mistakes in. I know many professional mathematicians who, at some point in their careers, have constructed a faulty proof, usually before publishing it. As soon as a colleague pointed out the difficulty, nothing prevented my friends from feeling miffed—but everything prevented them from objecting. As soon as the point was understood, they'd say something like, "Oh, damn!" Mathematics is that clear.

One case of bad deductive science concerned a "proof" of the famed four-color theorem by Alfred Bray Kempe, an English barrister. He published his proof in the nineteenth century and everyone thought the matter had been settled until P. J. Heawood, a British mathematician, caught the error in the 1920s. The hunt for a proof was on again, and it culminated in the 1970s with the discovery of a correct proof. The proof had so many special cases that it took a computer to enumerate them! In this most prominent of examples, however, there was no controversy. At first, mathematicians believed that a theorem had been found; then they believed a theorem had not been found; then they believed again that a theorem had been found. It apparently has. No one ever argued about anything.

To include the deductive side of science in my compendium of bad twentieth-century science, I must turn from misapplied methods to misapplied math. There are many valid theorems that describe the capabilities of neural networks, those cybernetic wonders so frequently touted by the media in recent years. But none of the theorems justify the wild claims made by those who would apply neural nets to any and every problem. In fact, from the point of view of the theory of computation (a branch of deductive science in its own right) neural nets are extremely weak as computational devices.

Since this book deals mainly with inductive science gone wrong, it is time to pass through the third door to make a third crucial distinction.

Getting Ideas and Testing Them

Behind the third door, a naked old man sits in a tub, merrily scrubbing himself. A bar of Ivory soap floats (naturally) in the water beside him. The old man stops scrubbing, stares intently at the soap for a moment, then suddenly yells

EUREKA!

It is quite deafening. The man is Archimedes, and the word means "I have it!"

Archimedes had been pondering the laws that govern floating objects. Speculating in his mind as he scrubbed himself, he suddenly had an idea: Is it possible that the weight of water displaced by the floating bar of soap actually equals the weight of the bar itself? Knowing only the deductive tradition, Archimedes is somewhat at a loss to answer his question. He has nevertheless visualized the matter and feels convinced. Luckily, Bacon enters the room from the previous chamber.

"What meaneth this crye?"

After I explain the matter to Bacon, he runs back into the other room, reemerging moments later with a weighing balance in one hand and a glass jar in the other. It looks as though he's going to apply the scientific method. Here's the apparatus Bacon assembles.

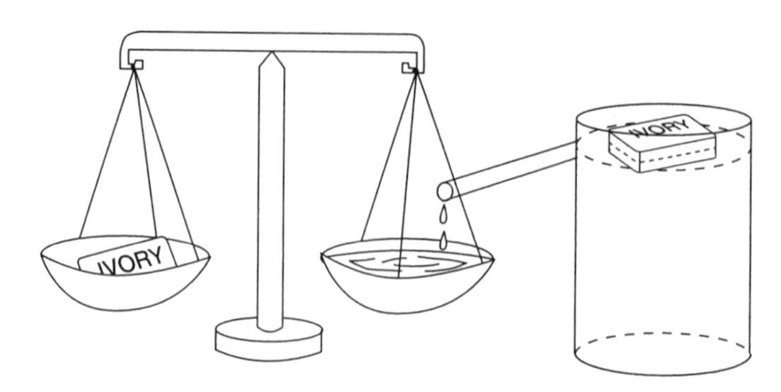

Since Archimedes has two identical bars of Ivory soap, Bacon puts one in the left-hand pan of the balance. He puts the other bar in the jar after channeling its overflow directly into the right-hand pan of the balance. When the water stops trickling into the pan, the balance seesaws back and forth a little, then settles into a horizontal position. Archimedes, who grasps the significance of the experiment at once, murmurs "Eureka" under his breath.

"Mayhap we were but luckye," says Bacon. He repeats the experiment.

We leave these old sorcerers now to their strange lives in the castle of science and turn to

The Scientific Method

The third distinction clarifies the scientific method, which applies only to the inductive sciences. Every scientific discovery has two parts:

1. Getting an idea (like Archimedes in the tub)
2. Testing the idea (like Bacon and the balance)

Formally speaking, we call any idea worthy of testing an hypothesis. It makes no difference how a scientist obtains an hypothesis. In search of a new idea, a scientist may plunge into a hot bath, ponder the universe in an opium den, visit a psychiatrist, lie on the floor kicking and screaming, whatever works. I'm not suggesting that scientists typically use more than one of these methods at a time, only that the process by which a scientist gets an idea worth testing is not, strictly speaking, part of the scientific method. In many ways it is the fun part of science, where researchers speculate freely on the laws that operate behind the scenes.

To establish an hypothesis such as "every floating object displaces its own weight in water," a scientist must apply the scientific method. In essence, the scientific method is little more than an elaborate reality check, such as the experiment that Bacon performed on the floating soap. Properly applied, the method acts as a safeguard from error. Every piece of bad science in this book may be traced back to a particular error of method.

The most egregious error is to frame an hypothesis but not bother to check it at all. Sigmund Freud did precisely this and then went on to claim that the psychoanalytic method actually worked. Speculation flourishes in an environment that is free from the chilling effect of actual experiments.

Represented as a sequence of steps in the diagram below, the scientific method begins with a question. What nuclear processes account for the degree of hydrogen that we observe in the sun? How much of deciduous forest litter is recycled by chemical mineralization alone? What colors attract the most attention from average shoppers in a supermarket setting? What is the highest temperature at which superconductivity may exist? Is the universe a closed space?

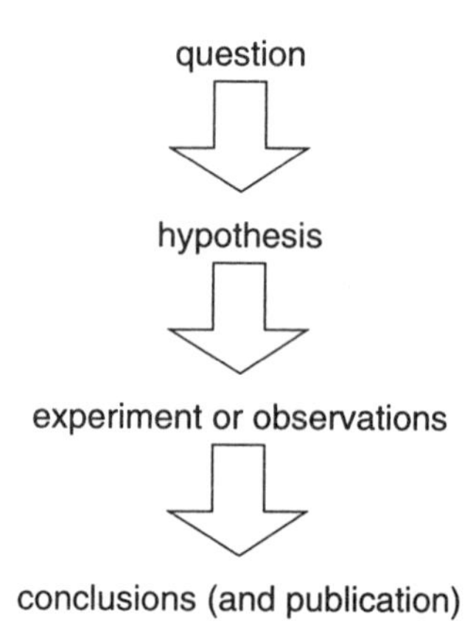

The question being asked strongly influences the course of the investigation. Woe to the scientist who asks the wrong question. And woe to the apprentice who doesn't even ask a question. The question, after all, must be scientific to the extent of addressing a general law, not a particular circumstance. To ask, Can human beings live inside a sealed glass house that contains a replicate biosphere? is not a scientific question because it involves no general laws. It is actually a technological question.

No question of any kind was asked at the conception of the intelligence quotient. Beginning as a measure of how fit certain underprivileged children were for school, it grew like topsy into an unwieldy social instrument that practitioners claimed could measure "intelligence." But since no one had ever defined intelligence to the satisfaction of more than a handful of people, it is simply impossible to say whether the IQ measures intelligence (whatever that is) or something else.

If the scientist asks a question, a scientific question, and then considers the question, an hypothesis may suggest itself. If the hypothesis can be tested, it will help to settle the question. In particular, an hypothesis must be falsifiable. The scientist must be able to imagine an experiment that will disconfirm the hypothesis if it happens to be false. Did the radio astronomer Frank Drake descend to the level of apprentice when he launched SETI, the Search for Extraterrestrial Intelligence? As I will argue in chapter 4, the SETI school seems to have dug itself into a nonfalsifiable hole. Although it continually fails to find any sign of extraterrestrial broadcasts, it claims that the search has not yet been conducted at an appropriate resolution or fineness of wavelength. There is no end to this game.

In the next step of the method, the scientist designs the experiment, performs it, and records what happened. The experiment must be well designed to help answer the question, settling the hypothesis one way or the other. Finally, the experiment must be repeatable. Nonrepeatablity was the major problem encountered by the cold fusion researchers, Fleishmann and Pons. In chapter 6, I describe how these two sorcerers-turned-apprentices could not repeat their own experiments with any degree of reliability: Sometimes they got anomalous currents from their electrolytic cells and sometimes they didn't. Other groups who attempted to repeat the experiments either failed to find anything or encountered the same on-again-off-again results.

This book begins with a very similar case that occurred at the beginning of this century. In 1901 the French scientist René Blondlot discovered (nonexistent) N rays. When other scientists tried to repeat these experiments, they failed.

As far as experiments go, an interesting difference arises between some of the inductive sciences. Physicists and chemists perform most of their experiments in laboratories, but some biologists and most astronomers do not have this luxury. They must substitute the observation of natural systems (whether storks or stars) for formal experiments. This substitution is reasonable if enough observations can be made to amount to an experimental program.

The experiment or observation may last months, even years, or it may be all over in a moment. But, in the end, the results must meet the goal of the experiment. They may confirm the scientist's expectations or deny them. In any case, they add to knowledge about the phenomenon under investigation. They may even call for a revision of the theory. As I will point out in the next section science is subject at all times to revisions at all levels, from small ones to large. All of these are "paradigm shifts" or none of them are.

It is not enough for the scientist simply to perform an experiment, however well conceived. He or she must communicate it to the rest of the scientific world. Strictly speaking, this is not a scientific requirement. A scientist may proceed utterly alone, refusing to divulge anything. This possibility raises the spectre of the mad scientist, subject of numerous Hollywood movies: "Deny me the Nobel Prize, will they? I'll show them."

To publish the results of experiments is both a social and a scientific act. Social because it enables others to share in the insights generated; scientific because publication enables other scientists to test the results. This acts as a further safeguard against error.

Science as a Fractal

Throughout the foregoing description, I have pretended that the scientific process is carried out by individual scientists, from beginning to end. This is not always true. Not only do many scientists work in teams on a specific research problem, but the individual members of a team may carry out only one part of the process, overall. Moreover, a great many more scientists constitute large, informal "teams," working on various aspects of a problem and communicating through correspondence, E-mail, telephone, and regular conferences devoted to the general area of the problem.

In addition, both deductive and inductive science have a fractal quality. They divide into subsciences and sub-subsciences, each with its own area of concern. This hierarchical structure reflects the structure of scientific problems. Although I have pretended that scientific problems have a simple, one-level structure, in general they do not. Many problems are part of larger problems while being divisible themselves into smaller ones.

This means that the framework of the scientific process must be understood to operate on several levels at once. A simple example concerns the geological question of how today's continents came into being. The discovery of continental drift has contributed to a solution of the larger problem but many subproblems remain: How was North America formed? Within the continent of North America, a geologist may also ask smaller questions, such as how the great inland seas of North America originated hundreds of millions of years ago. Within the time frame of one of these seas, the Tippecanoe, a geologist may study many sedimentary deposits in the attempt to discover the extent and duration of the sea. Another geologist might study just one deposit. The Ledyard Shale of western New York State contains a hidden record of temperatures and sedimentation rates for this sea. A paleontologist may study the trilobites of the Ledyard Shale. And on it goes, each study contributing to the whole.

Magic Formulas

The deductive sciences have supplied the notation, the tools, and in many cases the models that have become indispensable to the inductive sciences. Numbers, formulas, vectors, matrices, tensors, spaces, even precise logical thought itself, have their origin in mathematics. Where do you think the inductive sorcerers get their magic formulas?

To this extent, inductive science is suffused with a deductive component. The formula

$$PV = RT$$

relates the pressure (P), volume (V), and temperature (T) of a gas that is confined in an enclosed space. The letter R stands for a constant number. The formula encapsulates a million different statements about how this gas will behave. If the pressure T increases, for example, so does the right-hand side of the formula. But for equality to prevail, the left-hand side must also increase. In other words, either the pressure P or the volume V (or both) must increase in order to maintain equality.

It is sheer magic that the formula works to a high degree of accuracy under the conditions it was meant to apply. Why should this be? Other formulas are less accurate. Some are even statistical in nature. It is very easy to develop a new formula that appears to describe a general phenomenon and to become so mesmerized by its elegance that you never test the limits of its applicability. This sort of thing happens with increasing regularity as you traverse the spectrum of the sciences from hard to soft.

Traditionally, the hard sciences have incorporated a great deal of mathematics into their structure over the centuries, whereas the soft sciences have not. This lesson has not been lost on the social sciences, which have spent the last few decades piling up mathematical models in the mistaken belief that the more formulas they can crowd onto a page, the more "scientific" their work will become.

A statistical engine coughs and sputters behind the theory of racial differences proposed by psychologists Arthur Jensen and Phillipe Rushton. The theory has created public confusion (not to mention racial tension) and publication of *The Bell Curve* by Murray and Herrnstein has only compounded the problem, as you will see in chapter 8. The conclusions of these researchers may be repugnant to the great majority of social scientists, but the methods are hardly unfamiliar.

Bad Science

We come finally to the question, What is bad science? To begin with, whatever it means, bad science must be distinguished from fraudulent science. Apprentices, by definition, honestly believe that they are sorcerers. Fraudulent scientists (and would-be scientists) know that they are cheating when they fudge experiments, steal ideas, or make false claims. They are more like evil magicians than apprentices. They may be motivated by the same ambitions that drive apprentices, but they rightfully draw scorn and indignation upon themselves when found out. Having studied apprentices, I almost pity them. There, but for a certain grace, go I.

A preliminary definition might simply state that bad science happens when someone strays in a fatal way from the scientific method. This usually leads to results that are wrong at worst or grossly distorted versions of the truth at best. I might add that because scientists spend very little time consciously considering the scientific method (or whatever version of it they may formulate for themselves), they are prone to slips: observational lapses, misapplied formulas, and so on. It is only when they slip in a big way, "discovering" something that isn't true, that they slip fatally.

A kind of hell awaits such apprentices, especially when (1) the result is important, claiming media attention, and (2) the apprentice is too stubborn to admit an error or abuse of method. The stage is then set for scientific controversy. It remains only to see how the scientific process might be subverted in arriving at such a point.

The sorcerer's real power resides not in the atomic pile or the recombinant gene but in an ever-deepening understanding of the universe and its structure. The sorcerer cares about such understanding, moreover. Much that we call "talent" springs from this caring and from nothing else. What is the apprentice, then, but someone whose dreams of glory (or wealth) obscure or displace true caring?

Good science, after all, is real magic. It staggers the imagination to realize how many physical phenomena follow theories and formulas with an uncanny accuracy that has nothing to do with our wishes, our creative impulses, or anything else but reality itself. It boggles the mind further to find that phenomena predicted by theories and formulas turn out later to actually exist. Why should reality be like that? It's pure magic! But scientific magic only happens when scientists allow the logic of a well-conducted investigation to override personal hopes or fears about the outcome. If there are dreams of glory, real scientists keep them, trembling, in the background.

1

The Century Begins

The Rays That Never Were

The year 1895 was a momentous one for Wilhelm Conrad Röntgen, a fifty-one-year-old physicist at the University of Würzburg in Germany. It was a momentous year for German science, too. Röntgen had just discovered X rays.

The effect of this discovery on the public mind of the day is hard to grasp today. The media spun the tale well. Röntgen, a careful observer (and a true sorcerer), had benefited from an enormous stroke of luck: Applying a voltage to an evacuated glass vessel known as a Crookes ray tube, Röntgen noticed that a black line developed on some photographic paper lying nearby. Only light radiation would do this. Yet Röntgen had covered the Crookes tube with a shield of black paper (the better to see its ghostly contents). After a series of careful tests, Röntgen was convinced that he had a new form of radiation on his hands.

Not to be undone by a spurious observation, Röntgen spent several weeks confirming the phenomenon, testing different materials and noting that some absorbed the new rays better than others. Röntgen correctly surmised that the rays were a hitherto unknown form of light. He called them X rays because in mathematics the unknown is so often called *x*. When fully ready, he published a single paper.

The media went on to speculate (as the media will) about the effects of the new rays on society. Proper ladies were horrified to learn that persons in possession of an X-ray device could see right through walls to watch them undressing without their knowledge!

By the year 1900, news of the rays had reached every corner of the globe. Röntgen became famous. In 1901 Max Planck announced the quantization of energy. The prestige of German science, already great, became enormous. French science, on the other hand, seemed stuck in the doldrums. The land of Lavoisier, Carnot, Becquerel, Gay-Lussac, the Curies, and a host of other distinguished nineteenth-century scientists had produced no such soul-stirring results as the new century dawned. What was wrong?

Blondlot to the Rescue

René Blondlot, a highly respected French scientist at the University of Nancy, was as fascinated as anyone by reports of the new rays. Like many physicists of the day, Blondlot had been experimenting with X rays, probing their properties in the hope of contributing to the growing body of knowledge. Perhaps Blondlot also dreamed of finding a new ray of his own, stumbling as Röntgen had on a completely new phenomenon by the same phenomenal stroke of luck. Chance, it is said, favors the prepared mind. If good luck favors the sorcerer, bad luck bends over backward for the apprentice.

Blondlot was studying the polarization of X rays, searching for materials that would screen out all but the vibrations in a single plane. At the time of his discovery, he was using a very hot platinum filament enclosed in a sealed iron tube. A thin aluminum window allowed the radiation to escape into the laboratory where its properties could be tested.

It is not recorded precisely how chance first favored the unprepared mind of Blondlot, but he noticed some peculiar things about the radiation coming from his emitting apparatus. It seemed to increase the luminosity of a nearby gas flame. It also caused a screen painted with calcium sulfide, already dimly illuminated, to brighten perceptibly. Dare he hope? What would he call the new rays? The letter X was already taken, and to call them Y rays would seem to give them second place behind X rays. He settled on N for Nancy, the home of his university.

Experimenting with N rays, Blondlot soon discovered that many materials emitted N rays naturally, including iron and most metals. Wood, however, gave off no N rays whatever. If he wrapped a brick in black paper and left it out in the sunshine, it soon became an intense emitter of N rays and seemed capable of storing them for a long period of time. N rays had other strange properties, including an ability to enhance human vision, particularly when it came to seeing the N rays themselves! Blondlot had clearly stumbled on a discovery of the first magnitude, one that would establish his own name firmly in the front ranks of world science and restore the Académie des Sciences to its former glory.

N rays were obviously far more subtle and mysterious (and important) than X rays. Blondlot began to publish, completing some ten papers on the subject before the end of 1903. French colleagues quickly joined the scientific gold rush. The great Henri Becquerel published ten papers of his own; Charpentier discovered that human bodies gave off N rays. There were more papers as Zimmern and Broca joined in.

In Germany, meanwhile, the kaiser demanded to see the new rays, inviting Heinrich Rubens of the University of Berlin to perform a demonstration. Rubens tried mightily to produce the rays but failed on every attempt. His frustration would bear bitter fruit for Blondlot within a year.

When rays of ordinary light pass through a glass prism, they undergo refraction, some rays bending a little, others bending more. This process produces a spectrum in which the red frequencies of light bend the most and appear at one end of the spectrum. Blue light bends less, so it shows up at the other end of the spectrum. This much was known in Newton's time, over a hundred years earlier. Since then, physicists had discovered more exotic spectra, including the amazing discovery, a mere fifty years earlier, of spectra emitted by individual elements such as hydrogen. Such spectra had no rainbow, but a series of dark or light bands that would lead indirectly to Planck's announcement at the turn of the century.

Blondlot used lenses of aluminum to focus the new rays and prisms of aluminum to disperse them into spectra in the classic experiment going back to Newton. The figure below shows how an aluminum prism dispersed the rays onto a screen. Blondlot used a ruling mechanism consisting of a wheel and a worm gear that moved a phosphorescent thread across the spectrum by minute amounts in order to catch variations in the intensity of the N rays projected by the prism.

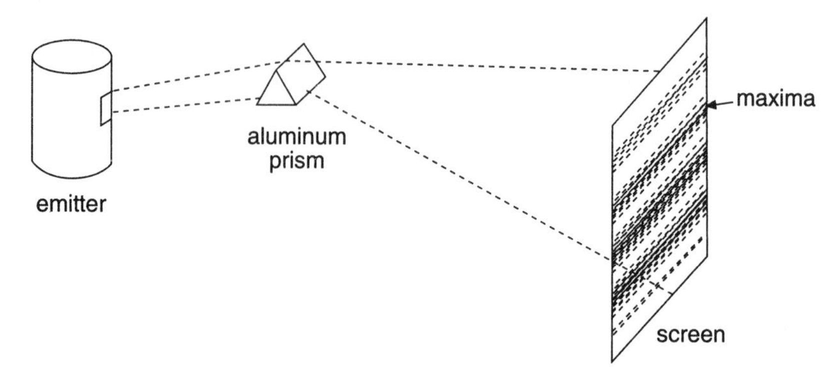

An N-ray spectrograph.

After emission by the hot filament in the steel tube, the N rays passed through a 2-millimeter-wide slit in an opaque material. This produced a narrow beam of rays that entered the prism. Blondlot must have been convinced of the wave nature of N rays to use such an apparatus. There was nothing wrong, in principle, in using a material other than glass.

Sitting in a darkened room, Blondlot would turn the wheel of his ruling mechanism, moving the phosphorescent thread across the spectrum, carefully noting each brightening and dimming of the thread. Detection of these light and dark bands in the spectrum whispered to the physicist of a very complex structure for N rays.

Wood Blocks N Rays

Following a meeting in 1904 of the British Association for the Advancement of Science, held at Cambridge University, a number of physicists gathered to discuss the N-ray phenomenon. Professor Rubens of Berlin, fresh from a frustrating attempt to reproduce the new rays for the kaiser, was particularly exercised on the subject. Someone would have to go to Nancy and test Blondlot's experiments directly. Since a European scientist might only aggravate sensitive national feelings, the task fell to an American physicist, Robert W. Wood.

Wood arrived at Blondlot's laboratory early one autumn morning. Blondlot, under the watchful eye of his assistant, drew the blinds, plunging the laboratory into darkness. He lit a small gas flame that would provide the faintest possible illumination.

Wood watched closely as Blondlot performed the first experiment. A continuous electric spark was placed behind a piece of ground glass to suffuse the light from the spark, making it easier to judge overall changes in the luminosity. Blondlot then focused some N-rays on the spark, immediately asking whether Wood could not see the spark grow brighter. Wood could not. Blondlot moved his hand to interrupt the beam of N rays, claiming that this dimmed the spark. Wood failed to see the effect. Blondlot sighed. Wood's eyes, he said, were clearly not sensitive enough. Wood then brightened. Perhaps Blondlot would be kind enough to call out the changes in brightness while he, Wood, moved his own hand into and out of the N-ray beam in a manner invisible to Blondlot.

"I suggested that the attempt be made to announce the exact moments at which I introduced my hand into the path of the rays, by observing the screen," wrote Wood in a report to the journal *Nature*. "In no case was a correct answer given, the screen being announced as bright and dark in alternation when my hand was held motionless in the path of the rays."

Blondlot next showed Wood some photographs that demonstrated the brightening of the spark. Wood found these unconvincing because the plates had been exposed cumulatively and an experimenter could unconsciously favor the N-ray exposure by holding the plate before the spark for a slightly longer period of time when the N-ray emitter was turned "on."

Frustrated, Blondlot suggested a demonstration of the vision-enhancing effects of his mysterious new rays. Wood enthusiastically agreed, and he watched as Blondlot held a flat iron file just above his eyes. Blondlot said that he could plainly see the hands of a faintly illuminated clock that rested on a table on the far side of the laboratory. He was enabled to do this because the file, like many metals, emitted N rays, which of course enhanced his vision. Wood asked whether he could hold the file. Blondlot agreed.

Wood had noticed a wooden ruler lying nearby and, in the darkness of the laboratory, substituted it for the file. Yes, Blondlot could still see the hands of the clock quite clearly—this in spite of the fact that wood was supposed to emit no N rays at all!

By this time, the assistant was casting hostile glances Wood's way. Blondlot suggested they visit an adjoining room where he had set up his spectrograph, an apparatus that used an aluminum prism to disperse the N rays onto a screen in the form of a continuous spectrum (see the description above). Blondlot positioned himself at the screen and began to turn the wheel, calling out the positions at which he saw bright or dark lines. Wood sat nearby, right beside the aluminum prism, watching skeptically in the dark.

As Blondlot called out the positions at which he read the maxima, Wood realized that they were barely a tenth of a millimeter apart. Yet the aperture through which the N rays passed on their way to the spectral band was over 2 millimeters wide. In ordinary physics, this would be impossible. A pinhole camera, for example, could never produce such definition with a 2-millimeter hole. All parts of the image would be smeared out by 2 millimeters or more, producing a very blurred picture. He raised the matter with Blondlot there in the darkness.

"That is one of the inexplicable and astounding properties of these rays," said Blondlot. He offered Wood a seat at the micrometer. Wood carefully moved the phosphorescent mark through the spectrum. He could see no variation whatever in the brightness of the mark. He retired to his own seat with a sigh. Perhaps Blondlot could try again.

In the darkness, lit only by a single gas flame, Wood glanced over at Blondlot's assistant who by now was glaring suspiciously at him. He waited until the assistant looked away for a moment, then deftly removed the aluminum prism from the bench. He waited.

Blondlot continued to call out the maxima in the spectrum. Wood said nothing. Before Blondlot opened the heavy curtains, Wood replaced the prism, but not before the assistant saw him adjust the prism to its former position. He rushed forward, placed the prism just so, glared at Wood, and suggested that he, the assistant, be allowed to read off the maxima once again. That way, the rude American might be convinced that the phenomenon was independent of the observer. Secretly, the assistant suspected that Wood had removed the prism. No doubt he would do so again. He would show the American who was clever and who was not.

He waited until Wood, with deliberately heavy footfalls, had returned to his seat and then began the search for maxima. *Mon Dieu!* There were no readings at all. Perhaps the American had made some *dérangement.* The assistant's ploy, based on the premise that Wood had once again removed the prism, failed. Wood, knowing that the assistant suspected him, had not removed the prism this time!

Fall of a Sorcerer

Within days, Wood filed a report of his visit to Blondlot's laboratory in the journal *Nature.* The appearance of Wood's article sealed Blondlot's fate and the fate of his beloved N rays.

On reading the article, the French physicist Le Bel declared, "What a spectacle for French science when one of its distinguished savants measures the position of spectrum lines while the prism reposes in the pocket of his American colleague!" But the Académie des Sciences, loath to relinquish the dream of N rays, awarded a prize and a medal to Blondlot in the December meeting of 1904. Both, they said, were given in recognition of Blondlot's work as a whole.

Papers on the subject of N rays continued to appear for a few years, then petered away to nothing. Blondlot, it is said, entered a period of severe depression from which he never recovered. Blondlot, the physicist who was the first to measure the conduction velocity of electricity in wires, eventually became mad and died as a direct result of the debacle.

French science recovered, of course. In 1900, for example, Paul Villard had already discovered gamma rays. It just took a few years for the significance of these rays, as indicators of nuclear transformation, to be fully appreciated.

What Went Wrong?

There can be little doubt that Blondlot was the victim of self-deception. Those with little stake in the outcome—German scientists, for example—not only failed to replicate Blondlot's experiments, but devised others that proved the nonexistence of N rays. Why hadn't Blondlot himself thought of these definitive experiments?

The scientists who knew Blondlot were universal in asserting his honesty and sincerity. Either of two hypotheses might account for Blondlot's deception. One hypothesis is straightforward; the other, twisted and dark.

When you think you've made a discovery of Earth-shattering importance, you must be made of the sternest stuff imaginable to devise the very experiments that threaten to turn the dream to ashes. Otherwise, you will tend to work unconsciously in a way that disguises the nonexistence of "your" phenomenon.

Blondlot could easily have deceived himself by merely working in a very dark room where visual threshold phenomena take over. Half-seen things are seen more (or less) clearly, depending on personal expectations.

The darker hypothesis concerns Blondlot's assistant. The physicist Cuenot declared, "The whole discovery of N rays might have been initiated by an overzealous laboratory assistant who tried to make himself indispensable to his professor . . . assistants are not usually given to a scrupulous love of truth and have little aversion to rigging experiments; they are quite ready to flatter their superiors by presenting them with results that agree with their a priori notions."

Was Blondlot the apprentice, or was his assistant? We may never know.

From the point of view of scientific method, Blondlot had produced an unrepeatable experiment. Nothing in physics rules out subtle phenomena per se, and Blondlot knew this as well as anybody. But any scientist faced with an observation this subtle must work all the harder to establish it beyond doubt. The pathos of Wood's visit might have been avoided had Blondlot simply asked himself, How can I prove that this phenomenon does *not* exist? He could easily have invited another scientist to study his gas flame as he, Blondlot, passed his hand into and out of the path of the putative N rays. Without telling the other what he should expect, Blondlot might have dealt himself a useful dose of frustration.

If Blondlot deceived himself near the beginning of the twentieth century, two other researchers would fool themselves near the end of it. The famous "discovery" of cold fusion by Martin Fleishmann and Stanley Pons in 1989 would eerily echo Blondlot's nonreproducible experiments (chapter 6).

Blondlot's Blunder Triggers Langmuir's Laws

Blondlot's blunder forms the centerpiece of a colloquium delivered in 1953 by Irving Langmuir, an American physicist and Nobel laureate. Langmuir had made a study of the bad science up to the 1950s and formulated its salient characteristics in the form of several "laws." They apply with special force to N rays and cold fusion. By "effect," Langmuir means a phenomenon that can be detected or measured, or seemingly so, by apprentices of one stripe or another.

Langmuir's Laws

1. The maximum effect that is observed is produced by a causative agent of barely detectable intensity, and the

magnitude of the effect is substantially independent of the intensity of the cause.

2. The effect is of a magnitude that remains close to the limit of detectability, or many measurements are necessary because of the very low statistical significance of the results.
3. There are claims of great accuracy.
4. Fantastic theories contrary to experience are suggested.
5. Criticisms are met by ad hoc excuses thought up on the spur of the moment.
6. The ratio of supporters to critics rises to somewhere near 50 percent and then falls gradually to zero.

As evidence of his laws, Langmuir cited not only N rays but the Allison Effect and the Davis and Barnes Experiment. (See the bibliography.)

You may enjoy pondering the chapters that follow, to see how many of the signs of bad science can be found in each example. Most examples display some of the symptoms; a few display all of them.

2

Mind Numbers

The Curious Theory of the Intelligence Quotient

In 1904, the very same year that American physicist Robert Wood made his fateful visit to the laboratory of René Blondlot, a countryman of Blondlot's by the name of Alfred Binet was asked by the French Ministry of Education to develop a test that would help to identify students with learning problems. This simple request would lead, by a circuitous route, to a piece of bad science that stood in stark contrast to the N-ray debacle then coming to a head. The history of the intelligence quotient began not with an apprentice but a true sorcerer, so to speak. And if the N-ray phenomenon was short lived, the IQ concept was to die only the slowest of deaths. It is, after all, still alive!

Binet, who directed the psychology laboratory of the Sorbonne in Paris, had long been interested in the scientific study of the human intellect. Years earlier, he had followed with fascination the craniometric studies of another famous countryman, Paul Broca (after whom Broca's Area of the human brain is named), who claimed that more intelligent people had larger heads. Determined to test this idea for himself, Binet visited several schools. After the teacher had identified the brightest and the dullest students in each class he visited, Binet measured the heads of these students, assiduously following the techniques recommended by Broca. Binet found the results sufficiently discouraging to abandon the idea of physical measurements altogether. The average difference between the brightest students and the dullest came to about a millimeter. Moreover, individual numbers varied so widely that some dull students had larger heads than some bright ones. The method was clearly not useful in determining the intellectual future of individuals.

Binet's honesty as a scientist cannot be doubted. As Stephen Jay Gould points out in his book *The Mismeasure of Man*, Binet even tested himself for bias in an independent head-measuring experiment. Horrified to discover that his own expectations had a perceptible effect on the measurements, Binet was all the keener to abandon head measurements altogether.

The request from the Ministry of Education gave Binet the opportunity to try a new, more inherently psychological approach to the problem. He devised a test that resembled an examination but which did not address scholastic questions. Instead, the questions on this test reflected a student's ability to reason about simple things such as coins, faces, and other everyday objects. He strove mightily to include as many different kinds of reasoning as he could think of, from counting, ordering, understanding, imagining, and correcting (errors).

By 1905 Binet had completed the first version of his test, in which he arranged the tasks in order of difficulty. In the second version of his test, completed in 1908, Binet rearranged the questions in order of "mental age." For each question, he reasoned, there would be a minimum age at which a normal or average child might reasonably be expected to answer it correctly. The mental age assigned to a student taking the test would be the age level associated with the last question that the child answered successfully before running into trouble. To ensure that a student's ability to read or write had no influence on the outcome, Binet administered the test orally by a direct interview.

The number that Binet attached to each student taking his test was the difference between the student's chronological age and his or her mental age, as revealed by the test. The German psychologist William Stern argued that Binet should not take the difference between these ages, but the quotient. If one divided the mental age (as revealed in the test) by the student's chronological age, one would have a quotient. Thus was the Q of IQ born.

As if aware of how his test might be later abused, Binet gave specific warnings about the dangers of misuse: "The scale, properly speaking, does not permit the measure of the intelligence, because intellectual qualities are not superposable, and therefore cannot be measured as linear surfaces are measured."

What Binet feared most of all was the process Gould calls "reification," a word we may translate as "thingifying." Just because we have a name in our mind does not mean that something specific or real has been named. For some people the name "unicorn" suffices to thingify a white, horselike creature with a single horn sprouting from its forehead. The name "unicorn" is certainly real, but the thing (as far as I know) is not. Binet also understood a second danger to lurk behind his test. The number it produced might well not be used as a guide to identify which students needed help, but a category to which they would be damned. Binet regarded intelligence not as a fixed quality or quantity, but one that could grow under the right tutelage. On the basis of special classes that he had designed and taught, Binet had no doubt that intelligence could increase: "It is in this practical sense, the only one accessible to us, that we say that the intelligence of these children has been increased. We have increased what constitutes the intelligence of a pupil: the capacity to learn and to assimilate instruction."

We leave Binet, happily applying his tests, identifying students that needed help and abjuring the use of the word "intelligence" in connection with them. Across the Atlantic, his tests were about to experience a curious rebirth at the hands of two apprentices.

The "I" Is Born

The intelligence quotient received its first public boost on this side of the Atlantic around 1910. H. H. Goddard, a director at the Vineland Training School for Feeble-Minded Girls and Boys in

New Jersey, found in Binet's test the ideal vehicle for making a most important distinction. In Goddard's time, psychologists defined "idiots" as those who never developed full speech and could barely progress beyond the general competence of a three-year-old. The next higher classification, "imbeciles," could speak well enough but seemed incapable of learning to read or write. An imbecile, by definition, had a mental age of somewhere between three and seven years. Early in the twentieth century, the words "idiot" and "imbecile" were almost technical in their use and content. Gradually, they would take on an increasingly pejorative character.

Goddard's "school" was not for idiots or imbeciles, but for the "feebleminded." To bring the taxonomy of mental retardation up to date, Goddard coined the word "moron." One level above imbeciles, morons occupied a gray area between idiots and imbeciles on the one hand, and fully competent people on the other. Morons might learn to read and write, but their skills would always be somewhat marginal. Binet's new tests, he discovered, were just the thing to detect morons.

The eugenics movement, started by the statistician Francis Galton in England two decades earlier, had taken root in America. There was much concern in some quarters that if the feebleminded and moronic were allowed to breed and produce children, the population as a whole would become polluted with these undesirable genes. Idiots and imbeciles posed no such threat, since they appeared to have little interest in (or competence at) reproduction, but morons were another matter. Domestic morons could be dealt with either by sterilization or by isolation (as in Goddard's school). But clearly, the arrival of new morons on American shores was much easier to deal with. Detect them and send them back to wherever they came from. In 1912 Goddard was commissioned by the U.S. Public Health Service to test incoming immigrants at the infamous Ellis Island facility.

Enthusiastically applying the Binet test to immigrants who could barely speak English and who were for the most part scared witless, Goddard arrived at some frightening figures. He found that 87 percent of Russian immigrants, 83 percent of Jews, 80 percent of the Hungarians, and 79 percent of the Italians (among others) were feebleminded. Deportations began to spiral wildly as a result of these tests. The number of deportations increased fourfold in 1913 and then sixfold again in 1914.

As Goddard would himself admit much later, the Binet test had its limitations. Perhaps it was wrong to deport so many people as a result of their scores on the tests.

The enormous expanse of the Atlantic crossed by immigrants had also been crossed by Binet's papers and copies of his tests. In the hands of H. H. Goddard they became precisely what Binet had feared. Goddard, after all, believed in "intelligence" as a single, fixed entity that could be measured more or less precisely. He also believed that it was passed on by a specific gene, one from each parent. Those who received no genes for intelligence would be morons, or worse. Those who received only one gene would be fit for "dull labor" but little else. Binet's test enabled Goddard to develop an intelligence quotient for each person tested. It quickly revealed the state of the testee's genes. Binet's remarks on the inapplicability of his test as a determinant of "intelligence" either did not reach Goddard or fell on deaf ears. Perhaps the hypothesis "IQ test measure intelligence" would never have occurred to Goddard for the same reason that no testing of such a hypothesis would be required, in any case. The thing was self-evident, after all.

Problems with the Binet scale and its application led Lewis M. Terman, an educational psychologist at Stanford University, to revise the test, producing by 1917 what we now call the Stanford-Binet scale. Terman extended the number of questions from 54 to 90. Many of the new questions were for "superior adults." While the Binet test had been administered orally by a trained tester, the new Stanford-Binet test was to be a written one. The new test, moreover, would hardly be confined to selected students. Terman already foresaw a universal IQ test: "What pupils shall be tested? The answer is All."

The Stanford-Binet test would be the foundation for all the tests to follow: Yerkes's Army Alpha and Beta tests, the Wechsler Adult Intelligence Scale, the California Test of Mental Maturity, the Cognitive Abilities Test, the Lorge-Thorndike Intelligence Test, the Otis-Lennon Mental Ability Test, and many others.

The Birth of *g*

About the time that Binet was commissioned by the French Ministry of Education to compose his famous test, English statistician Charles Spearman invented factor analysis, a technique for teasing out underlying uniformities in large numbers of correlations. To understand what this means, let me backtrack for a moment.

Two measurements, repeated a large number of times, may show a high correlation, a low one, or one that is negative. If I measure the lengths of arms and legs for a great many people, I will find a high degree of correlation (nearly 1.0) between the two sets of measurements. There is a simple reason for this particular correlation: People with longer legs frequently have longer arms, too. They are bigger and the relative sizes of the many people measured will be the main reason for the high correlation.

But if I measure the lengths of their arms and their hair (suitably defined), I will probably get a negative correlation. That is because people with longer arms would tend to be men, and men (even today) tend to have shorter hair than women. It is doubtful that the correlation would get very close to −1.0, the theoretical lower limit for the correlation of any two measurements.

Many positive (and, equally, many negative) correlations are entirely spurious. For example, I might find a very high correlation between daily stock prices and temperatures from March to August. But perhaps the market was going steadily upward just as winter was turning into summer in the Northern Hemisphere. There might well be a complete absence of any causal relationship between the two sets of numbers, yet they would show a high correlation. The omnipresent danger of thingifying rears its ugly head again.

Even if some plausible relationship might be imagined to hold between two highly correlated sets of measurements, the causal relationship can only be suggested by the high correlation, never proved. Mathematics is silent on the subject of cause when there is a high correlation, but mathematics has much to say when there is a low or zero correlation. When two measurements show a near-zero correlation, we can be pretty sure that the variables being measured are unrelated by any causal connection.

Spearman, who treaded where angels evidently feared to, sought a way to tease out causal effects by submitting the correlations produced by many measurements to a special analysis that can be represented visually as the assignment of axes to a cloud of points. The figure below shows a hundred measurements of two variables represented as points in the plane. Each point represents a person, let us say, and the coordinates of the point are the two measurements made on that person.

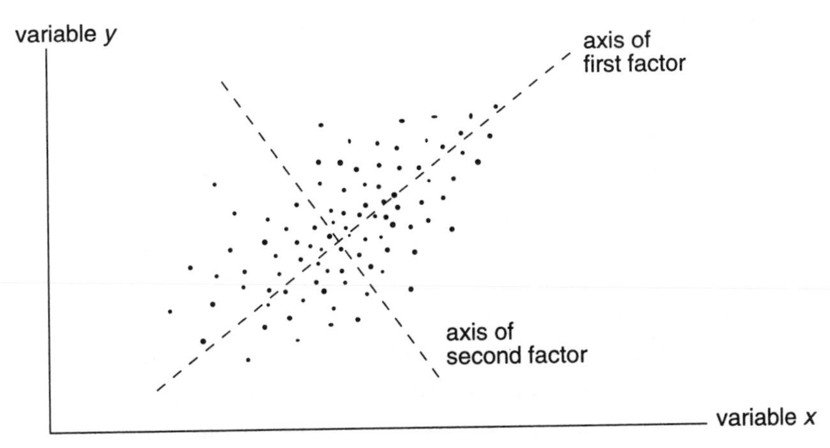

A pair of measurements is made for 100 people.

In this case, a positive correlation exists between the two measurements. The line slanting from lower left to upper right expresses this correlation. It may be called a principal component of the measurement data. It represents a "factor" that may (or may not) be present in the data themselves. A second line, at right angles to the first, expresses a secondary, negative correlation between the two measurements. It represents a second "factor" that operates independently of the main or principal factor.

When examining the data of IQ tests, Spearman was struck by the high degree of correlation between scores achieved by people who took two different tests. Was there some underlying factor common to the tests? To find out, he applied factor analysis and discovered that, indeed, there was. He called it *g*. Spearman meant this letter to stand for "general intelligence," a perfect example of thingifying. Because he already believed in a thing called general intelligence, an innate quality, Spearman was perfectly happy to assign a real status to this abstract (and, all too often, meaningless) factor.

In fact, Spearman was ready to declare that through his discovery of *g*, psychology would attain the status of physics as a hard science. Spearman claimed a universal character for his *g*, not to mention top place for himself as a scientist of note: "All branches of intellectual activity have in common one fundamental function. . . . This *g*, far from being confined to some small set of abilities whose intercorrelations have actually been measured and drawn up in some particular table, may enter into all abilities, whatsoever."

Spearman, who longed for the kind of perfect, abstract generality that had previously belonged only to mathematics and physics, saw in *g* the salvation of psychology, not to mention his own elevation into the front ranks of scientific genius. Although the charge of cargo-cult science lay still far in the future, Spearman would undoubtedly assume it applied only to the statistical vagaries of his plodding colleagues. However, beyond claiming a special role for *g* in all human intellective processes, Spearman was hard put to identify any specific mechanisms or to throw any light on just what "intelligence" was. He was in the position of a fisherman who felt an enormous weight on his line and declared, "I have a whale here." But was it a whale or a wharf piling?

The lack of a solid hypothesis involving an alternative definition of intelligence, as well as the total absence of any experiments designed to test the hypothesis has been amply demonstrated. It remains to wonder just what "intelligence" is and to evaluate IQ tests as an assessment of it.

What Is Intelligence?

For the foregoing reasons and others as well, the IQ school has been under more or less continuous attack from the beginning. The concept of IQ has been criticized by psychologists, biologists, physicists, mathematicians, and philosophers of science. To counter these criticisms, the IQ school has cleverly drawn its intellectual wagons into a circle. In the words of Edmund Boring, a Harvard psychologist and early tester, "Intelligence as a measurable capacity must at the start be defined as the capacity to do well in an intelligence test. Intelligence is what the tests test."

This statement is a perfectly acceptable definition of something, but not necessarily of intelligence. To the extent that he was a scientist, Boring would have been aware that one does not give preempted names to any measurable entity unless it can be established that the entity is the same one that the earlier use of the name refers to. In view of the fact that psychology has no well-accepted theory of intelligence to begin with, the name "intelligence" was not preempted in the sense. However, it had (and continues to have) a popular meaning as embodied in a variety of contexts in which people use the word. To insist on calling whatever it is that IQ tests measure "intelligence" is not only to beg a very large question implicitly, but to confuse or influence the popular meanings explicitly.

The continuing use of the term "intelligence testing" gives teachers, students, and all who are tested or involved in testing, the impression that intelligence is actually being measured. They come to believe that such tests reflect an innate quality. So much so, that as a result of an ordinary test a student might *earn* a high (or low) grade, but as the result of an intelligence test, he or she *has* a high (or low) IQ.

What a difference a word makes! Suppose that Boring had said "Gzernmplatz is what the tests test." Who would want to have their gzernmplatz tested? What educational authorities would care about the level of their students' gzernmplatz?

Boring's definition can only be construed as an admission of failure disguised as a triumph of reason. As many authors have pointed out, a statement of the form "*X* is what *X*-tests measure," may be respectable as an operational definition, but it closes the door to any other, completely different sort of test or measurement.

By a theory of intelligence, I mean a theory that defines intelligence as a quality that inheres to some degree in every compartment of human mental activity. At a minimum, such a theory would have to be capable of identifying intelligent behavior as observed in a variety of natural settings from social interactions to athletic performance to intellectual work. The fact that such a theory does not at the moment exist does not excuse the testers from proceeding in the complete absence of such a theory.

In a fascinating book called *The IQ Controversy,* authors N. J. Block and Gerald Dworkin offer the following requirements for a theory of intelligence:

> What should such a theory [of intelligence] do? It should explain the causal role of intelligence in phenomena in which intelligence has a causal role. For example, it should explain how intelligence affects learning, problem-solving, understanding, discovering, explaining, and so on. Also, it should explain how factors which affect intelligence do so. A good theory should say something about what intelligence is and what people who differ in intelligence differ in (information processing capacity? memory?), though this latter task would presumably be a byproduct of the former tasks.

One might well add that some people seem to show more intelligence in one area than another. For example, some people are excellent at calculating social relationships, but are quite lost when it comes to weights and measures. Some people see analogies between things almost instantly, but seem unable to imagine new situations. These are just a few of the compartments within which intelligence may operate. I will expand this list implicitly when we come to examine the tests themselves.

Computer scientists have striven mightily in the field called artificial intelligence to produce what might be called intelligent programs. After several decades of unstinting effort, most AI researchers have come away with a profound new respect for even the simplest mental functions. One researcher told me that even to imitate how her two-year-old son could pick M&Ms out of the grass was a feat beyond most vision systems. Programs that read and understand children's stories have met with only limited success. Attempts to implement logic in complex situations, such as those faced by humans on an everyday basis, have produced a bewildering array of theories and approaches, with little to show in the way of real progress.

It is only when it gets to crisp, well-defined situations with simple rules (the computer's forte), that artificial intelligence has met with any success. Programs that play challenging mental games such as chess and checkers have only recently begun to meet with success against the most able human challengers.

With a myriad of possibly quite different mental operations being comprised under the word "intelligence," we arrive at the prime complaint of most critics, as expressed by Walter Lippmann, editor of the *New Republic*, in 1922:

> Because the results are expressed in numbers, it is easy to make the mistake of thinking that the intelligence test is a measure like a foot rule or a pair of scales. . . . Provided the foot rule and the scales agree with the . . . standard foot and the standard pound in the Bureau of Standards at Washington, they can be used with confidence. But intelligence is not an abstraction like length and weight; it is an exceedingly complicated notion.

Taking the Test

Since Binet's time, intelligence tests have been composed in two steps: (1) dreaming up questions to include in the test, and (2) modifying the test so that, when applied to a large group of people, it produces scores that are highly correlated with the Stanford-Binet test or one that is equally respectable. The theory behind step (1) is simple: The Stanford-Binet test measures intelligence, so any test that produces the same distribution of scores must also measure intelligence.

The following categories of questions are among the many that have appeared in IQ tests over the years:

Category	Example
Odd man out	house igloo bungalow office hut
Complete the sequence	7 10 9 12 11 __
Anagrams	Who was not a famous composer? ZOTRAM SATSURS REVID MALESO
Visual analogies	Which figure is missing? a square above a circle a square to the right of a circle a square below a circle
Unscrambling sentences	"Whites their you until of fire eyes don't the see"
Complete the pictures	A pig without a tail A man without a nose A tennis court without a net
Maze tracing	A complicated maze, seen from above with a clearly marked entrance and exit

Such questions clearly beg a certain cultural familiarity that might well elude (or temporarily puzzle) a recently arrived immigrant, an inner-city child, or a laborer who has rarely held a pencil, much less written or drawn anything. In addition, IQ tests frequently include such "commonsense" items as the following:

Give two reasons why most people would rather have an automobile than a bicycle.

Why is it better to pay bills by check than by cash?

Why is it generally better to give money to an organized charity than it is to a street beggar?

These questions are clearly outside the experiences of many children in inner-city schools. Who could possibly prefer a check to cash, for example? One question even stated "When a dove begins to associate with crows, its feathers remain ______, but its heart grows black."

It can hardly be doubted, given the general nature of IQ tests, that they carry an inherent cultural bias that favors one socioeconomic class over another. IQ testers turn this very argument on its head, however, by claiming that high IQ scores correlate with "success," as measured by that very status. Authors Block and Dworkin have dissected this claim in the following interesting way: A typical correlation between IQ and success claimed by testers is 0.5, not a high correlation, but not one you can ignore, either.

How is the 0.5 correlation of IQ with success supposed to show that IQ tests measure intelligence? One argument (A) is

1. Success is a measure of intelligence
2. IQ correlates with success
3. Therefore IQ is a measure of intelligence

Representing the situation by a triangle with corners labeled Success, IQ, and Intelligence, Block and Dworkin label the lines of the triangle as shown in the picture below.

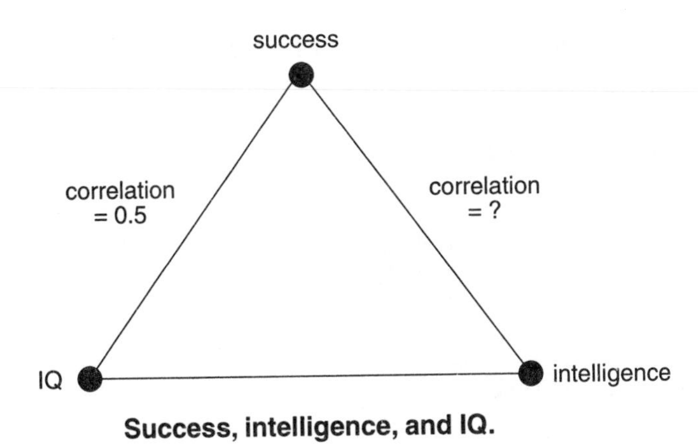

Success, intelligence, and IQ.

For the sake of argument, let us suppose that success is a perfect measure of intelligence, i.e., that the correlation of intelligence (as it would be measured by an ideally perfect intelligence test) with success is 1.0. Then, since IQ correlates 0.5 with success, it follows (deductively) that the correlation of IQ with intelligence is 0.5. This would

> make IQ tests a rather poor measure of intelligence, for it would mean that at most 25 percent of the variance in IQ scores is due to variation in intelligence. . . . So making premise 1 of A as strong as possible makes the argument as strong as possible (i.e., deductively valid) but only at the cost of making the conclusion rather weak.
>
> But how strong need the conclusion be? That is, how high must the correlation of IQ with intelligence be for it to be legitimate to say IQ tests measure intelligence? . . . We would require that IQ scores reflect mainly intelligence, for example, that somewhere in the vicinity of, say, three-quarters of the variance in intelligence can be predicted from IQ. This would correspond to a correlation between IQ and intelligence of 0.87.

Block and Dworkin then reverse the direction of the argument:

> We will now show that if the conclusion of A is interpreted very strongly, the premise can no longer provide much reason for believing it. Let us suppose that . . . the correlation of IQ with intelligence is 1.0. . . . it follows that intelligence correlates 0.5 with success. But then a person who propounds argument A and also accepts its conclusion is in the position of arguing that IQ is intelligence on the ground that IQ correlates 0.5 with something (success) which itself correlates only 0.5 with intelligence. That is, he is arguing that $x = y$ on the ground that both x and y correlate 0.5 with z.

Is IQ Inherited?

To the degree that IQ (as measured) turns out to be a highly plastic number, one cannot claim that it is inherited to any significant degree. Perhaps the most telling demonstration of the plasticity of IQ came in 1946 when Bernadine Schmidt, a young social scientist from Chicago, published a classic study in the journal *Psychological Monographs.* Schmidt's article, an unprecedented 144 pages long, described changes in the social, cultural, and intellectual behavior of 254 children of ages twelve to fourteen. The children, who all came from disadvantaged or dysfunctional homes in the Chicago area, had all been classified as "feebleminded." Their average IQ was 52, as compared with a nationwide average of about 100.

Schmidt conducted an intensive three-year training program that involved personal behavior, fundamental academic skills, manipulative arts, and good study habits. At the end of the period the students were tested again and proved to have an average IQ of 72, a full 20-point increase. Five years later, Schmidt tested her subjects again and found the average had increased to 89 with one-quarter of the students having gained more than 50 points. The study was exhaustive and impressive but never duplicated.

An Intelligent Ending

During this century, the notion of IQ as an innate quality, not to say a heritable one, has permeated the American collective unconscious. Throughout their checkered history, IQ tests have been used not only by educators anxious to measure academic potential but by those who would establish differences between races. I will revisit this particular question in chapter 8 in a review of the work of Arthur M. Jensen, Phillipe Rushton, and the authors of *The Bell Curve,* a work laden with cargo-cult formulas and other exciting developments.

These authors have raised storms of controversy over the years. Lately, political correctness has nearly succeeded in submerging the real scientific debate. Among all the criticism leveled at Jensen, Rushton, or the authors of *The Bell Curve,* only one man hit the nail on the head. It was Richard C. Lewontin, a well-known geneticist, who stood almost alone: The problem is IQ itself.

From Binet's day to the present time, no general theory of intelligence has emerged from the field of psychology. The state of our knowledge about this key phenomenon has remained in almost the same state for the last hundred years. It follows as a deduction that no one has ever confirmed that intelligence tests measure intelligence. It is even possible that research in this field has been inhibited by the belief that we already know what intelligence is.

At present, IQ tests are very much on the decline. One might hope that by the year 2000 we will hear no more of the intelligence quotient. We will still have tests, one hopes, to identify children who need help, but we will call them by the name that some testers have already begun to use. What IQ tests seem to measure is scholastic aptitude, and little else.

Is "Luck" Real?

Having been inspired by the methodology of IQ testing, I am ready to assert the independent reality of luck, a degree of which has been apportioned to every human being at birth. To assist me in this cause, feel free to make a copy of these pages, fill out the examination, and mail it to me care of the publisher. I will release the results to all who send in completed forms.

A Luck Quotient (LQ) Test

Instructions: You must answer all questions as honestly and accurately as you can.

1. (calibration) Who is the luckiest of the three people below? The person who
   (a) finds a $5 bill, (b) finds a $10 bill, (c) finds a $100 bill
2. How often do you find forgotten folding money in clothing you haven't worn for some time?
   (a) once a month, (b) once a year, (c) never
3. Have you ever been involved in a serious accident without being physically damaged?
   (a) yes (b) no

4. Do you seem to get promotions on your job without asking for them?
   (a) yes (b) no
5. Have you won more money in lotteries, in total, than you spent on them?
   (a) yes (b) no
6. (negative) Have workers ever accidentally dropped heavy tools on your head?
   (a) yes (b) no
7. Do people frequently tell you that they wish they had your luck?
   (a) yes (b) no
8. You worry about money:
   (a) a lot, (b) a bit, (c) not at all
9. (calibration) Who is the unluckiest of the three people below who go swimming? The one who encounters:
   (a) a life buoy, (b) a harbor buoy, (c) a great white shark
10. Have you ever been involved in a serious accident?
    (a) yes (b) no
11. Have you enjoyed reasonably good health so far in your life?
    (a) yes (b) no
12. Do you seem to feel sick or depressed far more often than you'd like?
    (a) yes (b) no
13. How many members of whatever sex interests you have ever expressed feelings of attraction toward you?
    (a) none, (b) two, (c) several, (d) a great many

If you also indicate your socioeconomic status, I will attempt a correlation between luck and success. I may even attempt a correlation between luck and intelligence.

3

Dreaming Up Theories

The Unconscious Con of Sigmund Freud

Eric Kandel, now one of the world's leading neurobiologists, began his training as a psychiatrist. In the 1950s an interest in the theories of Sigmund Freud took him to the New York College of Medicine where, as part of his training, he did rounds at Bellevue Hospital. Classmate Allen Silverstein recalls those rounds. "In my first year, as a prospective psychoanalyst, I used to try to attend as many psychiatric grand rounds as I could. On one, a young woman who was suffering from grand mal epileptic seizures lay there incontinent and biting her lip. As one after another psychiatrist stood up to say that her seizures were an exaggerated form of masturbation, I decided that this was definitely not for me."

Kandel also switched fields, ultimately to join the search for the physical basis for behavior in the humble sea slug *Aplysia,* a search that I shall rejoin at the end of this chapter. Ironically enough, Freud began his career of medical research by studying the biology (and nervous systems) of simple creatures.

Kandel and his classmate had experienced at first hand the medical legacy of Sigmund Freud, one of the most enigmatic and influential figures of the twentieth century. You can pick up the story of Sigmund Freud anywhere in his seventy-five-year life span and find worshipful tales told by his followers. Our story begins in October 1908 in Salzburg, Austria, site of the First International Psychoanalytic Congress. Freud and his disciples were under pressure to prove to a doubting world that Freudian psychotherapy worked. The psychoanalytic movement started by Freud three years earlier was growing steadily despite opposition from skeptical scientists, both inside and outside the psychological community. The time had come to publish at least one successful case to reassure followers and to silence skeptics.

At the congress Freud presented the now famous case of the Rat Man: Ernst Lanzer, an officer in the Austrian army. He suffered from a compulsive fear of rats. He also feared that a lady friend, as well as his father, would be eaten by rats.

Freud interpreted Lanzer's obsession with rats through word association. Lanzer's father, for example, was a *Spielratte,* or gambling addict. His lady friend, on the other hand, might someday marry Lanzer. The German word for marry, *hieraten,* also contained the word rat hidden inside. According to Freud, Lanzer also associated rats with children and even identified with rats. To Freud the whole thing was clear: Lanzer's secret wish was to have anal intercourse with both his father and with his lady friend. The repressed wish had spilled over into the obsessive fear.

Freud reported to the First International Psychoanalytic Congress that he had cured Lanzer by reconstructing his early childhood based on oedipal theory: Lanzer's father had interrupted Lanzer's normal sexual development as a child by threatening him with castration. Merely mentioning this to the patient resulted in "the complete restoration of the patient's personality."

To his followers, Freud's theory sounded entirely plausible, but others, surely, came away from the congress shaking their heads. They were powerless to stop Freud, unfortunately, for his theories carried a huge social benefit for whomever espoused them. Viennese liberal thinkers could look terribly chic by spouting Freud. It gave their words a shock value that quickly set them apart from bourgeois, straitlaced Austrian society. The craze had already begun to spread to America where the psychoanalytic movement quickly took root, permeating the entire North American culture by midcentury.

One can divide the scientific work of Sigmund Freud into two categories: the theories and the case histories. I will call the theories what they are—hypotheses. They make up almost the entire corpus of Freud's writings. In medical research, case histories can sometimes suggest new and promising areas of investigation, but they cannot replace well-designed experiments. As everyone knows (or should know), Freud never conducted a well-designed psychological experiment in his life; yet he claimed on many occasions that his work was "scientific" and a contribution to "science." I will return to the grand hypotheses later, because its many tangled currents mark the wake of Freud's thought through an ocean of conjecture.

The "Experiments"

Among the (non-Freudian) scholars who have analyzed Freud's case histories and hypotheses most closely are Adolf Grunbaum of the University of California at Riverside and Frank Sulloway of M.I.T. Sulloway, a historian of science and author of *Freud, Biologist of the Mind,* points out, among other things, that Lanzer's fear of rats grew out of the account by a fellow officer of a horrific Chinese torture in which the victim had a pot strapped to his naked posterior. According to this story, a typical barracks myth, the torturer had placed a large and hungry rat in the pot and would proceed to terrorize the animal with a red-hot poker introduced through a small hole in the bottom of the pot. The rat had only one way out, to gnaw its way through the patient's buttocks, possibly right into his anus.

According to Sulloway, Freud reported to the 1908 Psychoanalytic Congress that he had seen Lanzer over a period of eleven months. In reality, Lanzer had left therapy after only a few weeks, well before the reported cure. Moreover, as Freud himself admitted to disciple Carl Jung in 1909, Lanzer's "father complex" continued to trouble the patient. Besides this glaring inconsistency in Freud's reportage, he apparently also falsified several key features of the analysis in a direction that would support his contentions about the source of Lanzer's illness.

Such problems merely hint at the difficulties that plagued Freud as he balanced the freedom his expository methods required against the demands of a scientific method that insisted on proof of hypotheses. As they stand, Freud's theories rest on six published case studies. The first three cases are inconclusive and incomplete:

1. Dora, an eighteen-year-old woman who suffered from "hysteria," fled therapy after three months. She stated that her father was carrying on with the wife of a family friend while that same friend was paying unwanted attention to her. Her father encouraged his friend in this project, thought Dora, to deflect attention from his own dalliance. Freud was having none of it. Dora not only secretly desired her father's friend, but her father as well. She left the couch, objecting to Freud's aggressive insistence on his own interpretation of her case.

2. An unnamed woman whom Freud "treated" for homosexuality, never actually "improved." She terminated the sessions after a few weeks.

3. Little Hans suffered from a phobia of horses. Freud saw him just once, and the lad's father, a devout Freudian, analyzed the boy himself, concluding that he suffered from oedipal feelings. In a fit of common sense rare in five-year-olds, Little Hans tried to convince both his father and Freud that he had been frightened of horses ever since witnessing a carriage accident. Freud and the father gradually wore Little Hans down in an effort to get him to agree to the Freudian interpretation of his phobia.

The remaining cases had a longer duration and reveal more detail of Freud's psychoanalytic method. They include the case of the Rat Man, as already described, and two others:

4. Daniel Schreber, a magistrate whom Freud never met, wrote a memo about inexplicable feelings of suffocation and the conviction that he was turning into a woman. On the basis of this memo, Freud diagnosed Schreber as a delusional paranoid whose condition had been brought on by repressed homosexual feelings about his father and older brother.

Schreber's father, however, was well known as the inventor of devices for improving children's posture, usually involving rigid frameworks of metal rods and straps that constrained children into various positions. Freud was undoubtedly aware of the father's preoccupation with such frameworks but seemed oblivious to the possibility that the senior Schreber had used his son as a guinea pig for many of his devices.

At the time, Freud was particularly keen to prove that paranoia had its origin in latent homosexuality. It is true that Schreber had experienced a fantasy of being a woman having sexual intercourse, finding it "rather pleasant." Schreber's illness only got worse. He developed the conviction that God was turning him into a woman. Schreber was committed to an asylum for the incurably insane. He worked for years to secure his own release, finally winning his case in a German court. At no time did he acknowledge a cure of any kind, despite Freud's claim that Schreber had been partially cured by reconciling himself to homosexual fantasies.

5. Sergei Pankieff, the Wolf Man, whom Freud diagnosed as having an "obsessional neurosis," was treated by Freud for a period of four years. Freud's psychoanalysis began with a childhood dream of Pankieff's:

> I dreamt that it was night and that I was lying in my bed. . . . Suddenly the window opened of its own accord, and I was terrified to see that some white wolves were sitting on the big walnut tree in front of the window. . . . In great terror, evidently of being eaten up by the wolves, I screamed and woke up.

Freud concluded, on the basis of this dream alone, that the young Pankieff had accidentally interrupted his parents in the act of coitus. Freud's main clue in this reconstruction was the fact that the wolves in Pankieff's dream were white, symbolizing his parents' white underwear.

Interviewed by the Austrian journalist Karin Obholzer in the 1970s, many years after his miraculous cure at Freud's hands, the Wolf Man at eighty-six complained of having lived his entire life with the same problem. He was a compulsive brooder paralyzed by self-doubt. In reality, Freud had never convinced the Wolf Man of the correctness either of his diagnosis or of the dream interpretation, for that matter. Freud had not realized, for example, that in the Russia of Pankieff's day, children slept in their nanny's bedroom, not their parents'. Complained Pankieff, "It's all false. . . . In reality the whole thing looks like a catastrophe. I am in the same state as when I first came to Freud, and Freud is no more."

According to Grunbaum, Sulloway, and a host of other writers, the hypotheses of Sigmund Freud are in deep trouble. For one thing, these six cases constitute the entire bulk of Freud's published case histories. As I pointed out in my review of the scientific method, a well-planned experiment or series of experiments and the subsequent publication of the results form two of the most crucial steps in inductive science. Case histories are not experiments, although case histories might well suggest hypotheses, as they sometimes did to Freud. Case histories might also lend credence to one's theories, without actually establishing them.

In this light, there is all the more reason to publish convincing case histories, as many as possible. Not only did Freud fail to do this, he didn't even come close. Instead of a hundred case histories, say, with largely positive outcomes, Freud published only six and not one of them supports any of the theories he happened to be applying at the time!

Although Freud's hypotheses are easy to dispose of as "science," some fascinating questions remain, such as how Freud developed them and how they came to be accepted so widely as "theories." The chief factor behind the whole development of

Freud's theories calls forth the subtheme of this book: the apprentice dreams of glory.

Failure to establish a sound experimental program, much less a critical evaluation of his own methods, leaves open the question of whether the hypotheses are "correct" or not. Incorrect methods do not automatically mean incorrect ideas. But if Freud was essentially guessing at the deep structure of the human psyche, what were his chances of being right? Was Freud a con artist who, so caught up in the promise of his own ideas, unconsciously conned millions of followers in the twentieth century? One thing is certain: Freud was never short of ideas.

The Hypothesis

Up to about 1885, Freud was a young and ambitious doctor just completing a stint at the Vienna General Hospital. He was depressed much of the time as he reflected on his condition: Although he had become quite a good neuropathologist, he found the practice humdrum. He longed, as they say, to "hit a home run." His first attempt was nearly a disaster.

In 1884 he had heard of experiments with a new drug derived from the South American coca plant. Trying it on himself, he found that it miraculously cured his depression. It brought meaning back into his life, restored his energy to marvelous new levels, and even suggested a route to success: Get everybody using this stuff. The world would thank him for it.

Freud pressed cocaine on his friends and family, wrote papers about its marvelous therapeutic benefits, and experimented with sniffing it, injecting it, and even smoking it. He found that when he made it into a kind of cigar, the drug worked quite well. During these days, Freud would take a little cocaine before attending a social outing. It turned boring dinner parties into absorbing social events of great significance.

But after championing the drug for nearly two years, reports began to circulate within the European medical community of cases of severe cocaine addiction. Freud was in trouble.

He backpedaled, drew in his horns, and said little more about the drug, even claiming to have ceased using it himself. He would have to find a new vehicle.

In 1885 Freud applied for and was granted leave to study with the famous Professor Charcot at a teaching hospital in Paris known as the Salpêtrière. What Freud found there fascinated him. Doctor Charcot treated mental patients by hypnosis. He demonstrated that some of them had an extra personality, unknown to the patient. It was Freud's introduction to the unconscious mind. Writing about this time of his life, Freud said, "I received the profoundest impression of the possibility that there could be powerful mental processes which nevertheless remained hidden from the consciousness of man." This much he certainly got right, although he was hardly the first physician to have such thoughts.

Back in Vienna, Freud published a paper on hysteria that raised the ire of his former mentor and patron, Theodor Meynert. Freud had, in effect, invoked a nonphysical cause for hysteria, rooting the disease in the unconscious mind at a time when most physicians looked for physical causes of mental illness. One of the few other doctors who invoked the unconscious as the source of mental illness was a Viennese colleague of Freud's, Josef Breuer. As early as 1880 Breuer had already developed (and become famous in medical circles for) his "talking cure."

Breuer's favorite case was Anna O., who suffered from hysteria, a common complaint of the time. In a state of hysteria, she would become mentally "absent." By getting her to remember her absent states and what she experienced during them, Breuer claimed to have cured his patient by a process he called "catharsis." Freud was greatly intrigued by the case. Anna O., meanwhile, was not too impressed, and remained cool on the subject of psychoanalysis for the rest of her life.

Freud and Breuer worked together from 1885 onward until their relationship terminated under unhappy circumstances. Early on, they arrived at what Freud, sensibly enough, called a "working hypothesis." The two doctors hypothesized that all mental functions involved a "quota of affect or a sum of excitation—which possesses all the characteristics of a quantity (although we have no means of measuring it), which is capable of increase, diminution, displacement and discharge, and which is spread over the memory—traces of ideas somewhat as an electric charge is spread over the surface of a body." If too great a "charge" were to build up in a person, he or she would begin to display one symptom or another of mental illness. The therapy suggested by the working hypothesis centered on "discharging" the built-up psychic disturbances that led to mental illness.

Although Breuer and Freud had, by their own admission, no means of measuring this "quota of affect," they found the idea so intriguing as to be worth pursuing by whatever means at their disposal. They would, at the very least, continue to theorize and to test their theories on patients. Up to 1893, Breuer and Freud had worked very well together, but by the following year the two were becoming estranged. Freud wanted to emphasize sexuality, seeing it as the real "quota of affect" that, when thwarted, could wreak havoc with the human psychic balance. Breuer felt distinctly uncomfortable with the new direction, not, he argued, because he was a prude (as some Freud supporters have claimed) but because their position, already somewhat tenuous in scientific terms, would thereby become rather extreme. Breuer was reluctant to stake out a position on the basis of the extremely small number of cases the two had actually analyzed.

Undeterred, Freud proceeded to publish a paper that located the origins of psychoneuroses in sexual life. At about this time, Freud invented the seduction theory: Psychoneuroses were all the result of childhood sexual traumas perpetrated by adults or other, older, children. By the mid 1890s the estrangement between Freud and Breuer was complete. Not only Breuer but others in Viennese medical circles were becoming increasingly alarmed by Freud's obsession with sex.

The break with Breuer echoed the break with his own teacher, Meynert, and foreshadowed similar breaks in the years to come. The typical pattern involved a close working relationship, followed by increasing mistrust on Freud's part, a divergence of theory (or claims of same), followed by a final break in which the previous collaborator found himself frozen out of the cadre of Freud admirers, which all this time was steadily increasing. It is just possible that whenever Freud foresaw the possibility of having to share the credit for a famous discovery, he would begin to maneuver both himself and his theories so as to put a gulf between himself and the potential rival. Freud showed ample evidence, throughout his life, of a strong, sometimes extreme paranoia. At the same time, he worked incredibly hard, sometimes long into the night, shaping and reshaping his theories. His energy and ambition seemed boundless.

The years from 1885 to 1895 saw Freud's gradual conversion from a neurologist with a strong belief in the somatic basis for mental illness, to a believer in pure psychology. Whatever the physical basis for illness might be, Freud felt increasingly that a pure psychology was possible, one that would not depend on neurophysiological details. The year 1895 marked Freud's last attempt to lay a neurophysiological foundation for human psychology. He worked incessantly on what he called the Project for a Scientific Psychology. In this work, Freud hypothesized different sets of neurons that would handle perception, memory, and consciousness. He postulated functions for each group of neurons. In various combinations and interactions, the functions appeared to explain normal states such as wishing, judgment, defense, cognition, expecting, remembering, observing, criticizing, theorizing, as well as various pathologies such as hysteria and hallucinations. He drew suggestive little diagrams of interconnecting neurons, recalled his cases, and thought long and hard about the whole panoply of human behavior, both normal and abnormal.

In the end, Freud was rewarded with a vision of sorts: "One strenuous night last week when I was in the stage of painful discomfort in which my brain works best, the barriers suddenly lifted, the veils dropped, and it was possible to see from the details of neurosis all the way to the very conditioning of consciousness. Everything fell into place, the cogs meshed, and the thing really seemed to be a machine which in a moment would run of itself."

At first Freud regarded the project as a triumph of reasoning, but later he confessed to his close friend and associate Wilhelm Fliess that he had grave doubts about the whole project. This was a reasonable reaction, given the incredibly ambitious nature of the project and the almost complete ignorance of neuronal functioning by the science of the day.

Wilhelm Fliess was a curious character on the Viennese medical scene. On the one hand he was well known for his theory that human psychic and somatic responses followed a complex series of cycles based on periods of twenty-three and twenty-eight days. So impressed was Freud by this theory that he came to believe that the numbers foreshadowed his own death at the age of fifty-one (51 = 23 + 28). But Fliess was also well known as the originator of a theory of human infant sexuality. He maintained that sexual life begins not in the teens, but in infancy, an idea that Freud took up with great enthusiasm. By 1899 psychoanalytic theory, as defined by Freud, asserted that specific types of "infantile libidinal fixation" gave rise to specific neuroses in later life. Fliess also maintained that all human beings were basically bisexual, another idea that Freud embraced and incorporated into his steadily burgeoning theory.

By the close of the nineteenth century, Freud's hypotheses were becoming so top-heavy that even he was beginning to find it all a bit confusing. Of the fantasies and distortions of memory that lay (he thought) behind psychoneuroses, he writes,

> I am learning the rules which govern the formation of these structures, and the reasons why they are stronger than real memories, and have thus learned new things about the characteristics of processes in the unconscious. Side by side with these structures, perverse impulses arise, and the repression of these fantasies and impulses . . . give rise . . . to new motives for clinging to the illness.

As the century drew to a close, Freud was completing his first book, *On the Interpretation of Dreams.* At the same time, he was suffering from a host of psychosomatic illnesses and paranoid fantasies. Strangely, he began to blame Fliess for his problems, although he was never specific about just what Fliess had done to cause them. In the opinion of more than one historian of science, the real reason may have been that Freud now regarded Fliess as a rival (just as he would later break with Jung for much the same reason). The man whom Freud had once allowed to operate on his nose (for an undisclosed ailment) became Freud's most bitter enemy. Freud was not only jealous of Fliess but, after the break, attributed the theory of bisexuality to himself!

As the twentieth century dawned, Freud's "theory" of the unconscious grew increasingly elaborate. Although he occasionally scrapped one aspect or another of his hypothesis, his published outlines grew increasingly difficult to follow. Those who would practice the new therapy grew increasingly dependent on the master for authoritative pronouncements on specific questions of psychoanalytic technique.

Madness in His Method

In the physical sciences, self-deception will ultimately be checked by the appearance of contrary physical evidence. But in the new "science" of psychotherapy, there was no experimental regimen, or even an experiential check on Freud's rich imagination. Nor would it always be clear when a particular clinical finding would amount to such a contradiction. No one, not even among Freud's followers, was sure enough of what his hypothesis actually meant to begin checking it. In these circumstances, Freud found fertile ground for what amounted to a gigantic confidence game.

To say that it was an unconscious con may help to explain why it succeeded so brilliantly. The car salesman who believes in his vehicle will almost certainly sell more cars, other things being equal, than the salesman who entertains doubts. Freud's con, then, began with himself. He acknowledged the role of belief in the spread of his ideas as when he begged Jung, "My dear Jung, promise me never to abandon the sexual theory. That is the most essential thing of all. You see, we must make a dogma of it, an unshakeable bulwark." Freud's character called for an absolutist position from the start. As numerous quarrels with early colleagues show, Freud would brook no contradiction of his ideas or methods.

Freud protected his hypotheses against questions from colleagues and reserved their interpretation (if they ever had any) for himself. Those who questioned the hypotheses, asking what they really meant, or demanding some kind of scientific proof, were described as "resisting," Freud's favorite word of defense. It was a brilliant idea, turning accusations (or even innocent questions) into manifestations of psychological ill health on the part of accusers. Freud also carefully cultivated his public image as a "genius," which word, in the public mind of the time (and perhaps still), allowed him to say practically anything he liked. The fact that so many of Freud's ideas came from other sources, never acknowledged by him, was lost on followers and public alike.

Freud would brook no rivals in the development of his psychoanalytic theory. Almost everyone who tried to take a creative part in developing the theory—Jung, Adler, and many others—found themselves estranged and attacked by Freud. Only the group that surrounded Freud, agreeing with everything he said, would keep the ear of the master. Freud referred contemptuously to this group of yes-men as "the crapule." Freud's extreme paranoia, coupled with his grandiose self-image, made it impossible for him (or sympathetic biographers) to credit anyone else with ideas that have come to be viewed as essentially "Freudian."

The study of Freud the man will, I venture to say, ultimately overtake the study of his theories. Freud, who burned all his private papers in 1907, did so apparently to create a myth about himself, to shroud the past in mystery, and to leave nothing degrading or questionable to trouble future believers in his greatness.

Psychoanalysis and Psychiatry

To show that Freud's theories were not "scientific," I merely had to point out that Freud omitted two major steps in the scientific method, namely experimenting and publishing. Being true to the principles outlined in the introduction to this book, I hasten to point out that Freud's failure to establish (or even explain) his theories does not automatically doom them to falsehood. At the same time, one might seek a quasi-confirmation of the theories in the role they have played in improving the mental health of people who have undergone psychoanalysis.

The fact that many people claim to have been helped by psychoanalysis proves nothing, unfortunately, since many more people will also claim to have been helped by psychics. But no psychic has seriously claimed that fortune-telling is a "science" in the usual sense of that word. Freud, on the other hand, frequently spoke of psychoanalysis as a "science" and of his theories as "scientific." They simply are not.

We undoubtedly have unconscious minds. All kinds of mental processes, from bodily regulation right up to the solution of sophisticated problems, are handled by portions of the mind that we may have no conscious experience of whatever. Anyone who has ever dreamed has experienced the unconscious directly. This much we already knew. We also know that the young of any species of mammal takes direct cues about its future behavior from its environment and from its parents. People can be terribly damaged by events, even purely social or psychological ones, in childhood. This much we knew, as well. Any future theory of human behavior must base itself on more reliable evidence than Freud's theory did. And it must go considerably further, if not exactly in the same direction.

Today, few psychiatrists lend much credence to Freud's theories, but the field has split up, somewhat. Some psychiatrists have been reduced to plundering the pharmacopoeia for tranquilizers and other drugs to relieve symptoms of mental illness. The fact that drugs seem to control severe mental illnesses such as schizophrenia hints that some of these illnesses may have a purely somatic origin, after all. Other psychiatrists practice various newer forms of psychotherapy that, when all is said and done, are just as shaky as the Freudian hypothesis. Still other psychiatrists take an eclectic approach, sometimes just trying to be a good listener or counselor operating on the basis of common sense.

It must not be forgotten that doctors who choose to become psychiatrists have elected one of the most difficult, stressful, and thankless areas of medicine. The fact that we have no working theories will make little difference to the psychiatric profession as a whole for the simple reason that we will always seek help when confronted by the terrifying unknown. On the day when some nonconscious process makes its presence felt, suddenly surfacing in the form of paranoia, obsession, panic, paralysis, or even hallucinations, we would seek help even from psychics.

Where Is the Mind?

Eric Kandel is but one of many scientists who seek the basis of mental function in the brains of simple animals like the sea slug *Aplysia.* He has spent years on this animal, teasing out the function of some one thousand neurons that are always the same from one individual sea slug to the next. He has discovered a physical basis for behavior that, in some ways, goes beyond what others had found before him, in this and other simple animals.

Kandel's research, like that of hundreds of other neurophysiological scientists, adds to a steadily growing, but increasingly complicated and puzzling picture of truly tiny brains. The road that leads from invertebrate neurophysiology to a new theory of human mental function is a long and winding one. Will this line of research ultimately reveal the physical basis of human mental function? Will it lead to fabulous new therapies?

It may be that another hundred years must pass before a real picture of the human psyche emerges. As for Freud, although he believed that an independent psychology of the mind was possible, he never ceased believing in the physical basis of human mental function, much as Kandel does today. He just couldn't wait for the results.

4

Surfing the Cosmos
The Search for Extraterrestrial Intelligence

It was noon, April 8, 1960. The recently completed 85-foot radio telescope dish at Green Bank, West Virginia, had just lost the star Tau Ceti below the horizon. Steering motors hummed and the great dish swung grandly to the south along the horizon until, like a great ear, it listened to another star, Epsilon Eridani. Up in the control room, radio astronomer Frank Drake and his colleagues listened eagerly to sounds coming from a loudspeaker. The sounds enabled the astronomers to hear the signals being intercepted by the dish.

Gathered in the 85-foot parabolic surface, electromagnetic waves, some of them from Epsilon Eridani and some from much further away, reflected to the focus of the dish where a large cylindrical housing sheltered a precisely tuned amplifier. The signals from the amplifier were fed to a chart recorder in the control room and, of course, to the speaker. It was a propitious day, the dawn of Drake's dream of intercepting messages from an alien civilization.

Called Project Ozma, the dream reflected Drake's conviction that somewhere out there, alien intelligences were transmitting helpful messages to less developed civilizations or, failing that, were at least inadvertently broadcasting their radio and television programs. Given the air of anticipation that surrounded the inaugural evening of Project Ozma, Drake and his colleagues can perhaps be forgiven for what happened next:

> [S]carcely five minutes had passed before the whole system erupted. WHAM! A burst of noise shot out of the loudspeaker, the chart recorder started banging off the scale, and we were all jumping at once, wild with excitement. Now we had a signal—a strong, unique pulsed signal. Precisely what you'd expect from an extraterrestrial intelligence trying to attract attention.

To check that the source of the signal really was Epsilon Eridani, Drake had the telescope taken off the target. The sound disappeared, meaning that this star (or a planet near it) may actually have been the source. Unfortunately, when they returned the telescope to track the star, the noise had disappeared.

An even more significant incident followed on the heels of the first one: One of the telescope operators told a friend about the signal and the friend contacted a newspaper. Before Drake knew it, he was deluged by calls from the media demanding to know what had happened.

"Have you really detected an alien civilization?"

"We're not sure. There's no way to know."

This answer could not have been better calculated to raise curiosity about the incident still further, guaranteeing a great deal of publicity for Project Ozma. A better answer would have been, "As far as we know, the anomalous signal originated right here on Earth." Both responses are true, of course, but the second would have a more chilling effect on the media. Drake, after all, was no stranger to anomalous signals.

At the tender age of twenty-six, he had been observing the Pleiades star group when a new signal suddenly appeared on his chart recorder. Drake recalls:

> It was a strikingly regular signal—too regular, in fact, to be of natural origin. I had never seen it before, though I had repeated the spectrum measurement countless

> times. Now, all of a sudden, the spectrum had sprouted this strong added signal that looked unusual and surely of intelligent design. . . . I still can't adequately describe my emotions at that moment. I could barely breathe from excitement, and soon after my hair started to turn white.

Drake never succeeded in recapturing the signal and today suspects that it may have been a military aircraft.

Since the exciting early days of extraterrestrial probing, Project Ozma has been succeeded by SETI, the search for extraterrestrial intelligence. Sponsored by NASA, the National Aeronautics and Space Administration, the SETI project, along with similar schemes, has absorbed over a billion dollars in congressional appropriations. Is the money well spent? The project has had and continues to have many critics, but few have gone to the heart of the matter.

As I will show in a later section, the problem with SETI lies at the very beginning of the scientific method—the hypothesis. Not only is it unavoidably geocentric, it is essentially nonfalsifiable. There is a troublesome formula, moreover, that is supposed to make the hypothesis seem more reasonable. As I will also show, the formula is a two-edged sword that actually argues against the hypothesis.

At this writing, none of the SETI projects have revealed so much as a whisper of alien intelligence. This has not stopped Drake from going out on a limb. He seems eager to "prepare thinking adults for the outcome of the present search activity—the imminent detection of signals from an extraterrestrial civilization. This discovery, which I fully expect to witness before the year 2000, will profoundly change the world."

When confronted with the failure of SETI programs up to this point, Drake wisely opines, "Absence of evidence is not evidence of absence."

Scanning the Skies

Radio telescopes supplement ordinary optical telescopes by giving us a picture of the cosmos by the light of radio waves. I say "light" because radio waves are just another part of the great electromagnetic spectrum, which also includes light waves. The scale below shows the entire observable electromagnetic (EM) spectrum, from the shortest waves, the X rays, to the longest waves that belong to certain radio emissions. The scale is logarithmic. This means that each horizontal segment of scale embraces ten times the numerical range of the segment to its left. As you can see, the visible part of the EM spectrum occupies only a tiny portion of the whole.

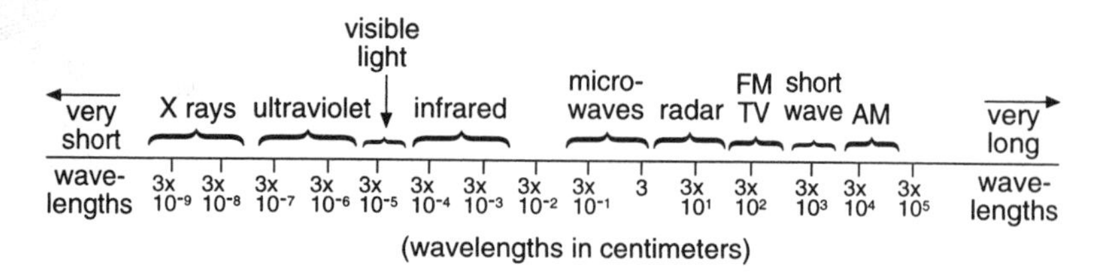

The central portion of the electromagnetic spectrum.
(Note that wavelength increases ten times with each hoizontal division to the right.)

EM radiation has two aspects: a wavelength that may be plotted as a position on the scale, and a frequency. Since all EM radiation travels at the same speed *c* (the speed of light), the frequency of the radiation is the inverse of its wavelength. In other words, if the wavelength of a signal is *w*, then the frequency is *c/w*, where *c* is the speed of light. As wavelength increases, frequency decreases. The wavelength of EM radiation is measured in units of length appropriate to the size of the waves, from nanometers to kilometers. The frequency of an electromagnetic wave, on the other hand, is measured in hertz (abbreviated Hz), which means the number of waves that pass a given point every second. Short waves with high frequencies might be measured in kilohertz (kHz), or thousands of vibrations per second; even shorter waves may be measured in megahertz (mHz), or millions of vibrations per second.

Because light waves are so much shorter than radio waves, they are capable of resolving distant sources quite easily. The picture of the cosmos revealed by optical telescopes is much sharper than the picture revealed by radio telescopes. In the first kind of picture, stars and galaxies stand out in crisp detail. In a radio telescope image of the same sky, there are only large, blurry patches that radio astronomers normally portray by means of contour charts.

Many of the stars that appear as sharp points in the light telescope map also show up as blobs in the radio telescope map. This means that such stars not only emit light, they emit radio waves. By the same token, some of the radio sources that show up in the radio maps have no visual counterparts, or if they do, turn out to be clouds of gas, regions of violent galactic activity. Radio telescopy has been an invaluable tool in learning more about the structure of our own galaxy and that of other galaxies, not to mention a string of amazing discoveries such as pulsars and quasi-stellar objects, or quasars.

Nevertheless, the relatively long length of radio waves makes it very difficult to get much resolution out of a single telescope. Indeed, the signal from a radio telescope is essentially one-dimensional, like a sound track. The signal from an optical telescope is, of course, two-dimensional—a picture. These days, radio astronomers squeeze more resolution from their instruments by using several receivers at widely separated locations, as if to construct an effective dish that has the same dimensions relative to radio waves that optical mirrors do in relation to light waves.

Consider a typical dish as it tracks a distant star. Radio waves pour onto the dish from all directions. Some of the radiation comes from the Earth itself, stray radio or television broadcasts, ham radio operators, taxi dispatchers, truckers on CB radios, cellular phone callers, direct-to-home-satellite broadcasts, and so on. These signals, the only evidence we have of intelligent life so far, all come from the planet Earth. They sometimes bedevil the life of normal radio astronomers.

Electromagnetic waves of more natural origin arrive from the ionosphere, where energetic particles from the Sun collide with molecules of air at the very top of the Earth's atmosphere. Electromagnetic waves also come from radio sources in our own solar system, such as Jupiter and the Sun. From beyond the solar system, faint waves arrive from other stars in our own galaxy, from pulsars, and from stellar clouds. Other ripples, ancient and feeble, arrive at the great dish from other galaxies, not to mention those primordial sources, the unimaginably remote quasars.

Life is nevertheless relatively easy for the normal radio astronomer who has techniques for eliminating many kinds of interference from earthly sources. He or she may listen largely undisturbed for the random hiss of ancient stars and galaxies or the repetitive clicking of a pulsar. But the radio astronomer who searches for intelligent life must stand this rationale on its head, listening for the whispers of intelligent transmission amid a welter of natural electromagnetic hisses, clicks, chirps, and buzzes. The SETI astronomer is even more bedeviled by earthly signals. They tend to sound just like the thing he or she is searching for.

Is it just possible that somewhere, among all the radiation flooding the dish from a myriad of sources, one or two indescribably faint signals amount to whispers from distant and ancient civilizations? Perhaps the signals patiently repeat the recipes for astounding scientific and technical breakthroughs, as some SETI enthusiasts have dreamed.

In the meantime, even as radiation from a multitude of sources pours onto the dish, a kind of reverse process goes on. All of those television and radio signals that interfere with the radiation from outer space are themselves zooming away from Earth in all directions at the speed of light. In their entirety, the signals form a vast, expanding ball of radiation. Since radio broadcasts began about ninety years ago, the radius of that ball of radiation is now about ninety light years. It is large enough to contain a few hundred stars in our near galactic neighborhood, albeit still tiny compared to our galaxy as a whole. It is nevertheless possible, of course, that a technological civilization on Alpha Centauri or Ophiuchus has picked up our broadcasts, including enough episodes of *The Three Stooges* to place the Earth in a state of permanent galactic quarantine.

The point of that rapidly expanding sphere of programming has certainly not escaped the SETI theorists. Other civilizations with the ability to monitor electromagnetic radiation should be able, sooner or later, to hear our signals, however faint. Should we not, by the same token, be able to pick up the signals of other civilizations? The prospect has an unquestioned fascination about it. Imagine what an alien signal might be like, the very stuff of science fiction! But is it science? Or is it fiction? In particular, is Drake behaving like an apprentice?

In this case, it all depends on the hypothesis and your opinion of it. As the well-known astronomer Carl Sagan once speculated, the sheer numbers of stars in our galaxy lends enormous weight to even the most slender estimates of probability for the evolution of technological civilizations elsewhere: What is the chance of Earth-like planets? What is the chance of life spontaneously emerging on such a planet? Even small probabilities, when multiplied by the enormous number of stars out there, turn into something almost definite. The hypothesis is this: Given the near-ubiquity of life in our galaxy, some life forms have surely developed intelligence, including the ability to communicate by radio waves. Such waves should be detectable by suitable receivers right here on Earth.

There are a few minor flaws in this hypothesis and one major one. The minor flaws involve unstated assumptions that enter the hypothesis. I deal with one of these minor flaws in the following section; the major flaw, at the end of this chapter.

The Apprentice Ponders the Cosmos

The introduction to this book outlined the scientific method as consisting of a few main steps. It also made clear that in "big science" in particular, a certain division of labor confines some scientists to just one of the steps. Some theoretical physicists, for example, do nothing but speculate about the ultimate structure of the cosmos. They use their imaginations to arrive at hypothetical models that will account for all the observed or suspected properties of the universe. For example, the current hypothetical consensus about the large-scale structure of the universe employs a mathematical model somewhat like a balloon. The universe is curved, according to this model, and will one day collapse back on itself.

The observational astronomers, who work the experimental side of the scientific method, may well cry out: "But where is the matter to effect this collapse?" If we estimate the mass of the known universe, we come up short by more than an order of magnitude. There does not appear to be enough matter to halt the outward expansion. Where is the missing matter?

The same theoreticians may propose that the universe contains many more neutrinos than we thought existed, or that an unknown "dark matter" lies plentifully scattered among the galaxies. Perhaps there is more hydrogen between the stars, or at the centers of galaxies, than we thought.

In most cases, observational astronomers have denied the proposals: No, we're not seeing nearly enough neutrinos. No, that much interstellar hydrogen is not detected. No, we already have enough matter in galaxies to account for their dynamical properties.

The hypotheses of the theoretical physicist or cosmologist "work" because the model is precise and one can tell, almost as soon as new observational evidence arrives, whether it confirms the model or not. If the speculation was far-fetched to begin with, the physicist should not be too surprised if further observations fail to confirm the model. It may be that the universe is not closed after all. The hypothesis must be falsifiable.

Consider now the scientist who looks up at the night sky and asks the age-old question, Is there anyone out there? The question seems perfectly reasonable. It means, Is there another race of beings, living somewhere else, whom we would call intelligent? Apart from the fact that we as yet have no formal scientific definition of intelligence (see chapter 2), most people think they know intelligence when they see it, at least among fellow human beings.

Perhaps the best laboratory in which to consider alien cultures is right here on Earth. Consider a country that is dominated by Zen Buddhism, for example. Many people would say that the Zen monk represents a very high level of human development (without being exactly sure what that means). If the world were full of Zen monks, however, we would be very unlikely to have radio. The technology would contribute very little to the insights necessary on the fivefold way, and one could argue that the technology and its development would constitute a completely unnecessary distraction from the real work of the monk, which is to rid himself of attachments to things of the world. As for advice from beyond, Zen monks have all they can handle in advice from the teacher.

With the peculiar myopia that characterizes western culture, we have come to regard our own development as more or less inevitable, an extension of the Darwinian imperative into the technocultural realm.

The real question is What is the chance of a Western-style scientific-technological civilization developing out there? The "Western" qualifier is crucial, for we in the Western world may be living in a spell, trapped in yet another aberrant vision of our place in the universe, one no less misleading than the pre-Copernican idea of a central Earth. If the sorcerer is under a spell, he will hardly do better than the apprentice!

Live by the Formula, Die by the Formula

The unquestioned pioneer of the SETI project is the well-respected radio astronomer Frank Drake. Early in his career as a radio astronomer, Drake developed an interest in the possibility of life on other planets, particularly intelligent life. He became intrigued, some might say obsessed, with the prospect of intelligent beings broadcasting radio signals into space, signals that we on Earth might intercept—to our infinite advantage.

Intuitively, Drake understood that with 200 billion stars in our galaxy, there might be a very good chance that someone out there was already sending the very signals he dreamed of receiving. To put the project on a quantitative footing, Drake devised the equation below. To some people it may appear complicated, but mathematically speaking, it could hardly be simpler. The right-hand side of the equation consists merely of a bunch of variables all multiplied together:

$$N = R^* \times Fp \times Ne \times Fl \times Fi \times Fc \times L$$

The equation attempts to estimate the number N of "radio civilizations" in our galaxy. A radio civilization is simply a race of intelligent beings that have developed the ability to broadcast and receive messages via electromagnetic radiation, and do so on a regular basis. The equation estimates the number N by taking into account a variety of factors in the product:

- R^* number of new stars that form in our galaxy each year
- Fp fraction of stars having planetary systems
- Ne average number of life-supporting planets per star
- Fl fraction of those planets on which life develops
- Fi fraction of life forms that become intelligent
- Fc fraction of intelligent beings that develop radio
- L average lifetime of a communicating civilization

At first glance, the equation seems perfectly definite. If you happen to know the value of each variable, you can come up with a pretty good estimate for N. If the estimate you arrive at is reasonably large, you may use the equation to squeeze endless amounts of money out of Congress to support a search for intelligent life. The equation, after all, is mathematical, and that means real science.

An estimate of the number R^* is based on an assumed rate of star formation of about ten a year. This is an extremely crude estimate based on current observations of regions where stars appear to be forming in our galaxy. The actual number has undoubtedly varied enormously over time, particularly in the remote past. For the Drake formula, it's all uphill from this point on.

The fraction Fp of stars having planetary systems is completely unknown. Although a few relatively nearby stars are suspected of having very large Jupiter-like companions or planets that are nearly stars in their own right, we have yet to observe a single star with a planetary system even remotely like our own. Period. It follows that we haven't the slightest idea what the real value of *Fp* might be, and any "estimates" would better be called wild guesses.

If we haven't a clue how many stars have planetary systems, then we're even more in the dark about the average number *Ne* of life-supporting planets per star. Some of them may well have such planets. Perhaps they all do. Perhaps our sun is the only such star. We simply have no idea.

Will a "life-supporting planet" ever develop life? I'm not sure how a planet could support life if it didn't already have it. The Earth has oxygen, for example, only because photosynthesizing organisms evolved here a long time ago and eventually filled the atmosphere with this (for us) vital gas. Perhaps the rather silly variable *Fl* should be set equal to 1 and simply dropped from the equation.

As you will see from a glance at the remaining variables, it gets worse.

The fraction *Fi* of life forms that become intelligent is even less well known, if that is possible, than the previous variables. What do we mean by "intelligent," anyway? As you may have already discovered in chapter 2, we're not even sure what we mean by our own "intelligence"! Once again, my guess for this variable is as good as Frank Drake's.

The fraction *Fc* of intelligent life forms that develop radio is likewise completely unknown and pointless to estimate. Finally, the lifetime *L* of the average radio civilization is the only variable about which we have any information, and that information may be about to improve. We know, for example, that our own radio civilization has existed for about ninety years. There is a real possibility that it may reach one hundred. In any case, this sample of one is our only basis for an estimation of *L*.

How do Drake and his disciples use the formula? Here are two examples that have appeared in popular magazine articles on the subject. I have no doubt that the guesses come directly from the SETI school.

$$N = 10 \times 0.3 \times 1 \times 0.1 \times 0.5 \times 0.5 \times 10^6$$
$$= 125{,}000$$
$$N = 10 \times 1 \times 1 \times 1 \times 0.01 \times 0.1 \times L$$
$$= 0.01 \times L$$

In the first guess, L was given a value of 10^6, or one million years. The second guess refused to assign a definite value to L, which is strange, considering that we already know more about L than the other variables. Nevertheless, using the value for L from the first equation in the second, we get a more conservative estimate:

$$N = 10{,}000$$

That's still quite a few radio civilizations. Why haven't we heard from any of them yet? We might find the answer by taking a closer look at ourselves, in particular, and our probable destiny as a radio civilization. It is not nuclear holocaust that will seal our fate as a spherical broadcaster of invaluable cultural and scientific information to the cosmos, but the incredible inefficiency of antenna broadcasting!

As every radio engineer knows, broadcasting electromagnetic waves in all directions at once is an enormously wasteful way to transmit information. Although the emissions from standard mast antennas can be directed somewhat in the form of lobes, only the tiniest fraction of broadcast energy ever reaches receiving antennas. The evidence is now very clear that the Earth is rapidly fading as a source of electromagnetic energy. Increasingly, we transmit radio and television signals by cable, not to mention the exponentially increasing Internet traffic on phone lines and fiber-optic cables. An even more powerful trend involves the broadcast of television signals toward the Earth from satellites, signals that are completely absorbed by the ground. The Earth may be about to vanish as a radio source.

If this is true, then 100 might be taken as a perfectly reasonable estimate for the crucial variable L. In this case, plugging the new value for L into the last equation at the top of this page, we get

$$N = 0.01 \times 100$$
$$= 1$$

That must be us.

Another implication of current developments in the dissemination of information points up another minor flaw in the Drake hypothesis. Increasingly, radio signals between points in deep space will be beamed ever more precisely at the target receivers, somewhat like a laser beam. This would make their reception by nontargeted civilizations increasingly less likely. Can anyone believe that these vastly "superior" alien civilizations would themselves employ any method so incredibly wasteful as spherical broadcasting to communicate with each other? The implications for SETI enthusiasts are clear: Don't hold your breath waiting for that magic signal.

Finally, there may well be radio signals that SETI will eventually intercept but the signals will present us with an enormous headache. Seemingly intelligent, they will only be meaningful to beings of a similar mindset, whatever that might mean. Neither I nor anyone in the SETI team can imagine what a distinctly inhuman mentality might be like.

Update on SETI

Harvard radio astronomer Paul Horowitz has directed the most sophisticated version of SETI to date. Called META (megachannel extraterrestrial assay), the survey used the Harvard/Smithsonian 26-meter dish in Harvard, Massachusetts. What made the survey sophisticated was not the dish but the radio equipment that scanned the incoming signal. There were, as the acronym hints, a lot of channels. In 1993 Horowitz concluded a five-year northern sky survey with the publication of a joint paper with Carl Sagan in *The Astrophysical Journal.*

In the paper, Horowitz described a search of the northern sky between –30° and +60° (most of the sky visible from Massachusetts) using a highly sophisticated receiver with over

8 million channels. Each channel embraced a very narrow range of frequencies, about 0.05 Hz in width. Such channels would be perfect to pick up the narrow-band signals that the SETI theorists expect other radio civilizations to use. Taken altogether, these channels cover more than 400,000 Hz, as shown in the diagram below. The frequencies center on the frequency of neutral hydrogen at 1,420 mHz (wavelength 10.5 cm). The diagram below shows the band surrounding this wavelength and shows, as well, other significant wavelengths. The chart is logarithmic, like the one on page 66 that showed the EM spectrum as a whole.

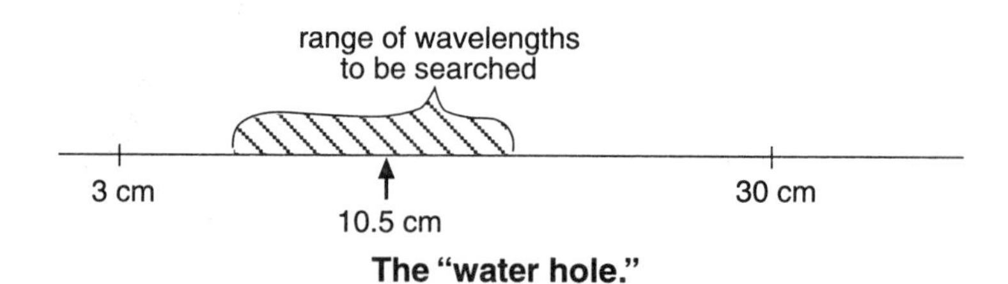

The "water hole."

Despite the high seriousness of the SETI mission, some of the scientists involved are not above a little humor. For example, they refer to the EM region around 1,420 (10.5 cm) as the "water hole." This frequency happens to occupy a particularly quiet stretch of the EM spectrum when it comes to interstellar signaling: the microwave region. Very near the frequency of neutral hydrogen (H), you will find the frequency of the hydroxyl radical (OH). Together, the two kinds of molecule form H_2O, hence the term water hole.

It must be said that Horowitz and Sagan were not searching for signals directly, but the carrier wave that might be modulated to form signals. This meant that if the search revealed reproducible bursts of energy in one or a few channels when the instrument was aimed in a specific direction, the astronomers would have no idea what the actual message looked like. They would then have to use a different type of amplifier to pick up the message on the carrier.

In their thorough search of the northern sky, Horowitz and Sagan discovered some thirty-seven anomalous signals that had the signature of a carrier wave. None were ever heard again, and the authors, in a fit of good-natured responsibility, dismissed them as probably not "signals of extraterrestrial origin."

But Horowitz and Sagan have also described plans to refine the search with BETA I, "the first high resolution, dual beam, all-sky search." Will BETA I succeed where META failed? And if BETA I fails, will the current proposers give up their search? No. BETA II already waits in the wings. And if BETA II fails, there can be GAMMA I, or what have you, with the rest of the Greek alphabet to work through. The point is that there exist a virtual infinity of refinements that SETI apprentices may try, an infinite succession of higher resolutions deployed in an infinite sea of possible search strategies. By definition, there will never come a point (short of depleting the congressional budget) where absence of evidence means evidence of absence. Speaking of the budget, Congress voted in 1994 to eliminate the allocation to NASA for its SETI program. This has sent SETI researchers scrambling for other means of support.

If the search program continues, it may only amount to a long and painful lesson in the dangers of nonfalsifiable hypotheses. Drake and the others may be in a trap of their own devising, unless they get very, very lucky.

But even if the SETI searchers succeed, triumphantly announcing contact with an alien civilization, this does not mean the science was good. Mickey, after all, did get the broom working for him (see Introduction). As you may recall, that was just the start of Mickey's troubles.

Personally, I expect the great announcement in 1999, just in time to save Frank Drake from his self-imposed prediction deadline. The signal will be highly compact and enciphered in some kind of code that no one will be able to crack. Then a brilliant graduate student at Cornell University will discover that the code amounts to an error-recovery system that protects the crucial content of the transmission from corruption during its long trip through interstellar space. The content, I expect, will turn out to be a two-dimensional scan signal, accompanied by a sound track.

Suddenly, we will at last have a truly privileged glimpse of some aliens themselves in the message picture. There will be three of them, perhaps short and yellow, one with a shaggy mane of curly hair, one with hair in the shape of an inverted bowl, and one with no hair at all. They will say "Neep, neep, neep," and jam their fingers in each other's eyes.

5

The Apprentice Builds a Brain

Misled by Metaphors

In 1962 a curious book called *The Principles of Neurodynamics* caught the attention of the computing world. Its author, Frank Rosenblatt of Cornell University, had created a series of machines called perceptrons that seemed to learn on their own. Built of a single layer of neuronlike elements, these machines could be trained to distinguish patterns presented to a small retina composed of simple on/off detectors.

Until his tragic and untimely death in a boating accident in the mid-1970s, Rosenblatt worked on his perceptrons and published claims that struck many computer scientists as too good to be true: ". . . The perceptron has established, beyond doubt, the feasibility and principle of non-human systems which may embody human cognitive functions at a level far beyond that which can be achieved through present-day automatons [i.e., digital computers]."

By the late 1960s a new generation of computer scientists was hatching all over North America. Graduate students stayed afterhours in computer labs to play with the emerging technology. Some developed new information-retrieval systems or circuit-design packages, but others saw a new frontier. New ideas from thinkers like Turing, von Neumann, and Wiener were in the air: robot software, self-reproducing machines, and cellular automatons. There were perceptrons as well.

On a thousand computer screens a thousand different images danced their way into the collective consciousness of the new generation. Some of the images, seen through a thin haze of strange-smelling smoke, seemed to herald a new age—a cybernetic age.

But in 1968 two M.I.T. professors, Marvin Minsky and Seymour Papert, spoiled some of the fun of perceptron enthusiasts by publishing a book called *Perceptrons*. The book, which pointed out a host of computational tasks that perceptrons could not perform, consisted of a large collection of theorems, each one proving yet another shortcoming. Perceptrons could not distinguish a connected figure (one that consisted of two separate pieces) from a connected one. Perceptrons could not do simple logic. The book had an immediate and chilling effect on the whole field.

But, like Mickey's enchanted broom, neural nets emerged from the perceptron splinters during the 1970s. The new networks had not one layer, but two, or even three. The new nets, it was said, had overcome the limitations that Minsky and Papert had so painstakingly described in their book.

By the 1980s a new generation of starry-eyed computer enthusiasts had emerged. The personal computer revolution had placed computational power that would have been unthinkable in the largest machines a decade earlier in the hands of millions of people, many of them in other sciences and technical professions. To some of the latter, the new neural nets sounded an irresistible siren song. Imagine a miniature brain that would do all your thinking for you. What else could a "neural network" be? Neural nets would not only solve hitherto unsolvable problems, they would select the best stocks to buy, help police departments decide which officers were honest, recognize faces and other patterns, solve scheduling problems, imitate human reasoning directly, and heaven knows what else.

The image of neural nets as miniature brains was pumped in the papers, science magazines, on public TV, and even in a few Hollywood movies. Neural network proponents (also called connectionists) did not rush to their phones to disavow these wild claims. In fact, John Hopfield of the California Institute of Technology claimed to have solved a most difficult mathematical problem called the traveling salesman problem: "Good solutions to this problem are collectively computed within an elapsed time of only a few neural time constants." No doubt many people believed such claims, and connectionists at the forefront probably enjoyed the publicity.

Neural nets are the brainchild of deductive science. They find their place as one computational strategy among many, some known, some as yet unknown. When it comes to cognitive processes (things that human minds do routinely), for example, there are symbolic reasoning systems that give neural nets a real run for their money. When it comes to pattern recognition, there are statistical classifier engines that outperform neural nets.

As an example of deductive science, there is nothing wrong either with neural nets or with the theorems that have been proved about them. These theorems have a general form that goes something like this: Given enough time on training examples, a given neural net will converge in its behavior to produce an approximate (or, in some cases, an exact) answer to a given problem.

But neural nets nevertheless illustrate what happens when people try to develop a whole technology based on a mistaken interpretation of what neural nets represent and what they can actually do. The next worse thing to bad mathematics is misapplied mathematics. As the saying goes, To a person who carries a hammer, everything looks like a nail. Are the connectionists like apprentices? They have done nothing to correct the impression that neural nets are all-powerful, intellectual hammers.

Imagine for a moment that the old sorcerer kept a book of methods and formulas compiled over the ages. Some of the formulas are very powerful and highly useful, others are mere curiosities, included in an almost recreational spirit. While the sorcerer slept one night, the apprentice leafed through the book. His mind boggled, quite understandably, when he came to the formula marked Neural Networks. Imagine, a brain in a box!

Aquiver with excitement, he constructed a neural net, tried it on a simple problem, and lo! it worked.

As computer recreations columnist for a number of years with *Scientific American* magazine, I had ample opportunity to review the literature on neural nets and may have unwittingly helped to spark the revolution by writing more than one article on the subject. Only now may I say what I never could in the column: The appearance of an idea in *Scientific American* does not automatically make it the coming thing. The number of people who seemed unable to spot the word "recreations" in my column was truly frightening. Although neural nets do solve a few toy problems, their powers of computation are so limited that I am surprised anyone takes them seriously as a general problem-solving tool. Most computer scientists who work in the field known as computational complexity understand this very well.

Neural nets have already entered the long, slow decline predicted by Irving Langmuir (see the end of chapter 1) for all bad science—or for that matter, technology.

What Is a Neural Net?

A neural net, as the name implies, is a network of neurons. The kind of network that most people have read about in recent years is the layered network, pictured in the figure at the top of the next page as a series of vertical columns of circles connected by horizontal lines. The circles represent neurons; the lines represent connections between neurons. The neurons on the left receive inputs from the outside world, and the neurons on the right make their outputs available to the outside world.

In between the input layer and the output layer, the typical neural net has one or more intermediate layers where most of the computation is done. Each layer, moreover, may have as many neurons as you'd like to put there. The neural network in the figure above, intended only as an illustration, is much smaller in these respects than most of the neural nets about which the world grew so excited in the 1980s.

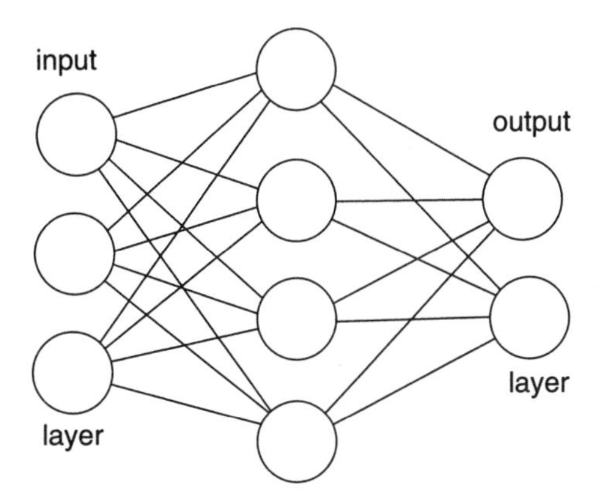

A simple neural net.

The operation of the net as a whole can only be appreciated by understanding how a single neuron operates. It receives inputs from all neurons to the left of it, each input a real number that may vary continuously through a range of values. You may think of these numbers as a kind of neural signal. As I will point out later, however, it isn't clear just what sort of signal the connectionists have in mind.

The neuron takes all the signals sent to it and adds them together. Then it submits the combined number to a sigmoidal function. This function does not get its name from someone called Sigmoid, but from the shape of the greek letter sigma. The diagram below shows the fate of various combined signals when they are processed by the sigmoidal function.

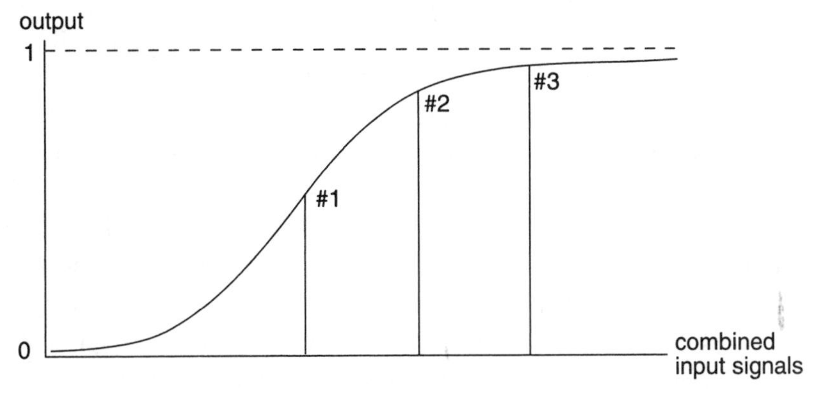

The sum of inputs is changed by the sigmoidal function.

The sigmoid shape, the great, sweeping S-curve that greets the combined signal lies between two lines, the 0-line below and the 1-line above. I have shown three possible values for the combined input signal along the 0-line or X-axis. Each signal is converted by the sigmoid function into a new signal that will be the neuron's output signal. The strength of the output is given by the height of the vertical line drawn from the X-axis to the curve itself.

The first of the combined signals, at position #1, meets the curve at a part where it is nearly linear, that is, close to a straight line. Within this region, a combined input signal turns into an output signal that stays pretty much in proportion to the input. The second of the combined signals, in position #2, meets the curve further along where it has begun a gentle, but never-ending approach to the 1-line. Here, because it no longer rises in step with the input, its response is called nonlinear.

The combined signal that arrives with a strength indicated by position #3 appears even further to the right, and the corresponding output is even closer to 1, but not much closer than the output for the signal at position #2. In this particular type of network, all neurons must output a signal that lies between 0 and 1 in strength.

Imagine now that someone sets the inputs of our neural net to specific values that collectively represent a problem to be solved, say the recognition of a specific pattern. The net goes through a convulsion of computation in which each neuron receives all the signals from neurons in the layer to the left, integrates these into a single output signal, and sends it to all the neurons in the next layer to the right. At the far end of the network, someone may record the output of each neuron in the final layer and this will be the "solution" to the problem posed at the input end.

I have oversimplified, of course. For one thing, I have left out the crucial component, the thing that makes neural nets so appealing to many people. Each line that connects a neuron in one layer to a neuron in the next layer has a kind of "synapse" that is capable of amplifying or weakening every signal that passes through it. In short, the synapse multiplies the signal passing through it by a numerical weight. The weights in a neural net can be changed, not by the net itself, but only by a special procedure that I will describe in a moment. In the meantime, we can alter the first diagram on page 83 by including these synaptic amplifiers, showing them as small triangles, one per line. See the diagram below.

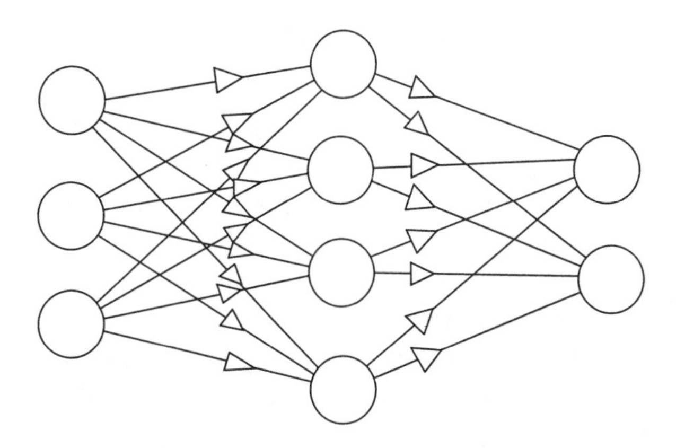

Synapses in the neural net.

Suppose now that the outputs do not quite amount to a solution of the original problem. There is a technique called back-propagation that adjusts the weights on each connection so that the modified network produces a set of output signals that corresponds more closely to a solution. And if that solution is not close enough, yet another round of back-propagation will bring it even closer.

There's a cute neural net about the size of the one displayed on page 83. It converts Cartesian coordinates to polar coordinates—more or less. The design was sent to me by Edward A. Rietmann and Robert C. Frye at AT&T Bell Laboratories in 1990. Cartesian coordinates are like coordinates on an ordinary, local map. Every point on the map has a horizontal coordinate (distance from the left edge) and a vertical coordinate (distance from the bottom). Polar coordinates also specify points with two numbers, but only one of the numbers represents a distance.

The other one represents an angle. Think of a circular radar screen on which every object's position may be described in terms of its distance from the center of the screen and the angle the distance line makes with a fixed, horizontal reference line.

With the right weights at all the network's synapses, you can feed any two numbers (Cartesian coordinates) you want to the input neurons, and in a flash the two output neurons will send out steady signals whose strengths yield the polar coordinates of the same point. It is a marvelous thing to see, and I don't mind that the network doesn't quite get all the transformations right. Most of them are pretty close. It would be small-minded of me to point out that the same transformations can be computed a thousand times faster by a ten-line program running on a standard microcomputer. After all, I only want to demonstrate the effect of the back-propagation algorithm and to probe its essential features.

The coordinate-converting neural network, or CCNet, begins life with all its weights set to arbitrary values, rather like your own synapses on a Monday morning. Then begins the education of CCNet. Training begins immediately with the presentation of a sequence of point coordinates at the input end and, for each of these, a comparison of the net's output with the true polar versions of the point in question. Each time there is a discrepancy, the back-propagation procedure adjusts the values of the synaptic weights, tweaking them up or down in a direction that improves the net's output for that particular input. Of course, the best synaptic values for the conversion of one point may not work as well for the conversion of another, but the ensuing adjustments balance out, in most cases, into an overall pattern of synaptic weights that together seem to compute all the appropriate conversions, no matter what points on the coordinate map are fed to it.

The CCNet, like the vast majority of neural nets, is not made out of hardware with artificial neurons, but from software. In short, most connectionists write programs that simulate a neural net, reproducing its intended behavior exactly. The synaptic weights, both before and after training, are stored in a table in the computer's memory.

Climbing in a Fog

The slight changes in the synaptic weights during the training period amount to baby steps through a multidimensional space in search of a peak. What does that sentence mean? Consider a mountain climber, shrouded in fog, standing at the base of a mountain range. Can he still climb the mountain? Of course. He only needs a hill-climbing algorithm. The climber, who can see (or feel) a bit of the ground around him, should always step in the direction of greatest upward slope.

To see where dimensions come into play, imagine that you are looking down on the climber (see diagram below) and can see two sets of direction lines, one east-west and the other north-south, superimposed on the climber's position. In general, the ground will slope upward along at least one of the cardinal directions. The climber may step in this direction, immediately increasing his altitude, however slightly.

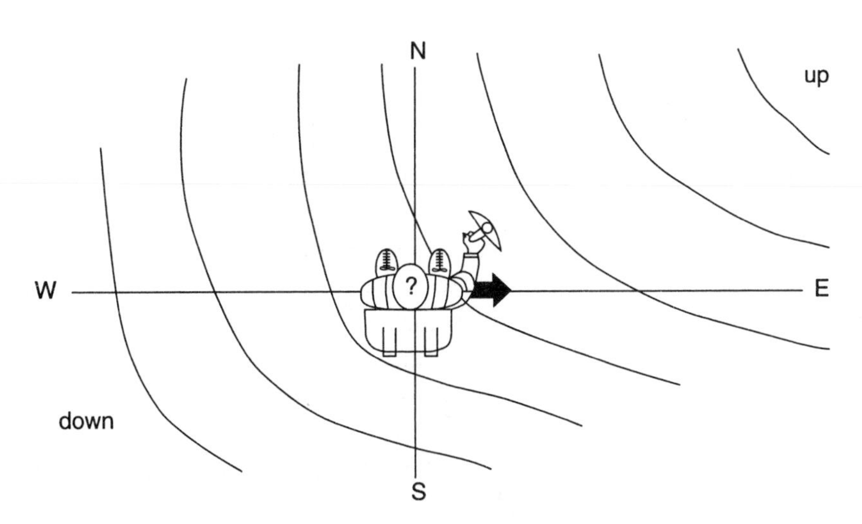

A hill-climbing algorithm in action.

The trouble with this technique is that the climber, after much weary toil through the fog, might well end up on a minor local peak instead of the grand snowswept monster he had aimed for.

Now, instead of a two-dimensional space, imagine the problem of climbing on terrain that is mapped on an n-dimensional space. Of course, this is a bit unfair of me. Nobody, not even mathematicians, can visualize a space of dimension higher than three. Nevertheless, the problem of finding a set of optimum synaptic weights can be thought of in these terms. Somewhere, there is an ideal set of synaptic weights (the top of Everest) that optimizes the response of the neural net over all possible inputs. With such a set of values firmly implanted in this artificial brain, it will produce the right answers in all cases or, at least, produce answers that are in some overall sense the best possible.

For each synapse there is a line along which you may imagine all possible synaptic values to be laid out in a continuum—like a ruler. If you set each synaptic line at right angles to all the others, you will have created the ground for an n-dimensional climber to walk upon. The peak being sought soars "upward" in the direction of an extra dimension, the n + 1st, if you will. To scale this extradimensional peak, the n-dimensional climber takes steps in the directions of greatest upward slope of the surrounding ground. Again, he may step along just one of the principal directions, changing only one synaptic value in a direction that improves the network's performance.

Even in this rarefied, multidimensional milieu, the algorithm may lead to a local peak in n-dimensional space. In fact, the more dimensions there are, other things being equal, the more opportunity a mountain range has for developing local peaks. This is not just a metaphor.

Mathematics is full of problems that amount to maximizing a multidimensional function. Most such functions tend to have many peaks, some higher, some lower. We already know that hill-climbing algorithms only work reliably toward a solution when there is but a single peak toward which all the ground slopes upward, which is certainly not the case in the illustration at the top of the next page.

The set of all possible combinations of synaptic values forms an n-dimensional space, and a particular problem to be solved will impose on that space a mountain range. At each particular point in the space, that is, at each particular combination

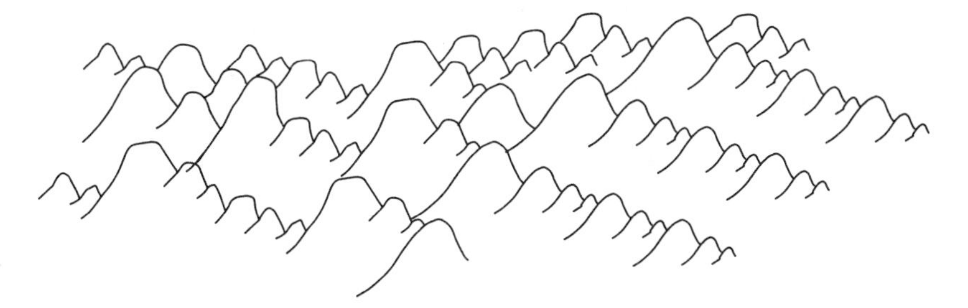

A difficult mathematical problem may have many peaks.

of synaptic weights, the fitness of the combination amounts to a height.

Two problems therefore confront the connectionist. First, the back-propagation algorithm is not guaranteed to produce a given network's optimum synaptic settings. Second, and far more serious, the optimum setting itself may be nowhere near what is needed for a truly useful solution to the problem at hand.

The *n*-Queens Network

The problem of placing *n* queens on a chessboard so that no two queens attack each other is well known to chess enthusiasts. Shown on the next page, for example, is a solution of the eight-queens problem on a standard chessboard. No queen on this board attacks another in the sense of being on the same rank, file, or diagonal.

In 1985 John Hopfield discovered a neural net that would solve this problem—or so he thought. In 1987 I was sent a copy of a program that simulated this neural net. I tested it on the four-queens problem and it worked. Then I tried the five-queens problem; it worked again. On the six-queens problem the neural net worked just once. Thereafter, on all higher values of *n*, the neural net failed to produce any solution at all, repeatedly placing queens on the same row or diagonal.

I was understandably skeptical when I came across the paper in which Hopfield claimed to have a neural net that could "solve" the famous traveling salesman problem: Given

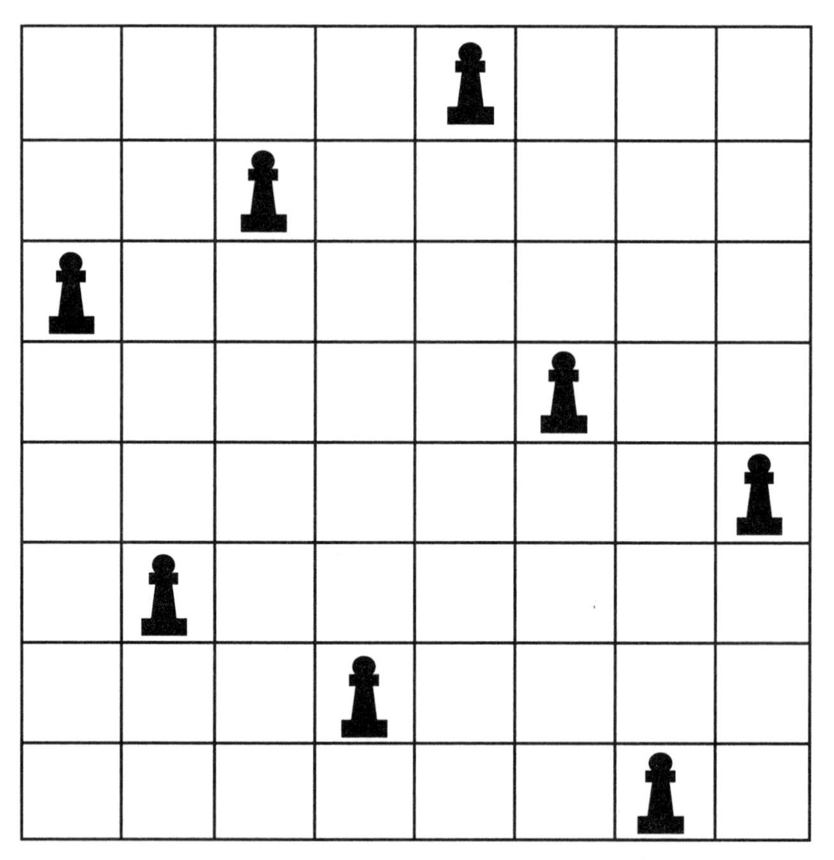

A solution to the eight-queens problem.

an arbitrary map containing cities and roads connecting them, devise a route that passes through each city once and has the minimum length over all such routes. The route is to be used by a hypothetical salesman who must not only visit all his potential clients, but must minimize gasoline costs. This problem is famous (at least among mathematicians and computer scientists) because it is notoriously difficult. No one has ever found a computer program that will solve it in a short time. In fact, the fastest programs known will find a minimal-cost tour in a time that is proportional to 2^n where n is the number of cities on the tour. Anyone familiar with the dread exponential function knows that if such a program took five minutes to solve a problem involving, say 100 cities, it would take ten minutes on 101 cities, 20 minutes on 102 cities, and so on. By the time it tackled 150 cities, the program would be taking the expected lifetime of the universe to find an optimum tour.

The traveling salesman problem belongs to a class of over a thousand problems that are all intractable in the same sense. The problems in this class share one other feature: If you could find a program that solves just one of them in a reasonable amount of time, you would have a program that solves *all* of them in a reasonable amount of time. This constitutes very good evidence that such problems are completely intractable in a practical sense. We can solve them, but only small examples. The rest are completely and forever out of reach for any computer, whether parallel, positronic, neural, or whatever.

I therefore suggested the project of testing Hopfield networks on a wide variety of problems to Gail Harris, a graduate student. Using Hopfield's own recipe for a neural net that was supposed to solve the traveling salesman problem, Ms. Harris found difficulties getting it to work on examples with more than five cities. Consultations with Hopfield proved of little value. He was vague about just how to set the net up properly. This forced Ms. Harris to find the optimal settings by a brute-force search through all possible settings. Even with these settings, the Hopfield network failed to find minimal tours on examples with eight cities and failed to come up with any sort of tour on examples with ten cities.

I learned that we were not alone in our research. G. V. Wilson and Stuart Pawley at the University of Edinburgh ran an experiment of their own on the Hopfield network and reported that it found valid tours on only 8 percent of the examples tested, and of these tours, none were minimal. In fact, their lengths were scarcely better than randomly selected tours. Researchers S. U. Hegde and J. L. Sweet at the University of Virginia also reported the failure of the Hopfield network on the traveling salesman problem.

This made me suspicious about the claims of the connectionists. They were beginning to sound like a cult, "voodoo science," as Zenon Pylyshyn, a cognitive scientist at Rutgers University, puts it. Perhaps the phrase "voodoo technology" would be more accurate.

Are Neural Nets "Neural"?

It can hardly be doubted that had neural nets been called something else, say "convergent networks," there would have been little or no public interest. It would have been about as popular as skiing on asphalt. But in today's whacky world of science-by-media, you merely need to give a scientific idea the right kind of name embodied in a catchy metaphor (like "neuron") to guarantee enormous publicity. Moreover, you don't have to provide more than a modicum of evidence that it actually works.

How "neural" are neural nets? The short answer is that neural nets have almost nothing to do with real neurons. First of all, neurophysiologists have yet to crack the code embodied by the rapid, stuttering pulses of energy that flow from neuron to neuron in real brains. Some neuroscientists believe that neurons use a frequency code to communicate numerical information with each other. The numerical signal transmitted from one neuron to another is the frequency of the pulses or spikes that it sends. The hypothesis certainly holds for signals to muscles for example; the higher the frequencies of pulses to a muscle, the harder it will contract.

Artificial neural networks transmit pure numbers, of course. The question thus arises, If real neurons transmit numbers in the form of frequencies, do they then add them together and submit them to a sigmoid function? Nothing of the kind. They don't even add the frequencies. Instead, they add the voltages that incoming signals produce on a neuron's membrane, and when the combined strength of these signals reaches a certain threshold, the neuron will send out a pulse of its own. The frequency of the outgoing pulses may bear little or no apparent relationship to the frequencies of the incoming ones.

It looks as though the apprentices mistook one kind of operation for another, adding frequencies when they should have been adding voltages.

There is another, deeper problem behind the neural metaphor: Much impressed by the analogy between computers and brains, some neurophysiologists have assumed that the neuron, like the electronic gate in a computer, is the fundamental unit of the computation we call "thought." The jury is still very much out on this one, as well. Why should an object as complicated as a living cell be the fundamental unit of any computation? For one thing, a cell all by itself may produce an amazing variety of behavior that can only be the result of still more fundamental units.

You only have to watch those free-living (single) cells called protozoa swim around in their natural world for an hour or two to realize that they make decisions about their course, when they will eat, and how they will move, all more or less continuously. I have personally seen one protozoan sneak up on another. Did it have a (subcellular) neural net that I couldn't see?

The neurophysiological parallel between natural and artificial networks vanishes almost completely under close inspection. All that remains is the tag "neural," like the grin on the Cheshire Cat.

The Synaptic Table

A neural net's prize, its Holy Grail, is the well-adjusted synaptic table. It contains the synaptic weights that may have been arrived at by hours or even days of computation as the back-propagation algorithm climbed slowly to the summit of a local peak in problem space.

Let's look at a hypothetical table: The entry at the intersection of row 385 and column 142 of the table represents the synaptic value of the connection between neuron 385 and neuron 142. The table is awfully big. It is common, for example, to have neural nets with, say, three layers with a hundred neurons per layer. This would mean ten thousand connections (and synaptic values) between neurons in the first two layers and another ten thousand connections between the middle and last layer. A table with twenty thousand entries is enormous and very difficult to analyze.

Let us indulge in a connectionist dream for a moment. Suppose that a large neural net was trained on numerically encoded descriptions of every disease syndrome known to humankind. Every time it made a wrong diagnosis, the synaptic weights would be readjusted by the back-propagation algorithm. Finally, the net's training is complete—and it works perfectly on every possible disease. Its synaptic table would then contain the boiled-down essence of diagnostic medicine, a great intellectual treasure. But how to get at the treasure? Of what value is an opaque, unreadable table? The knowledge embedded in that table would be inaccessible to science and therefore valueless as a scientific resource. Science is about accessible knowledge, not numerical hodge-podges.

The Challenge

The problem of finding the past tense of English verbs seems far removed from the demands of twentieth-century technology, but the problem is a favorite among experimental and theoretical psychologists and cognitive scientists as a testing ground for pet theories on how the brain works.

David E. Rumelhart and James L. McClelland are two prominent members of the PDP Research Group, an organization of connectionists spanning several universities and research institutions. In 1986 Rumelhart and McClelland developed a neural net that learned the past tense of English verbs—or seemed to. A number of cognitive scientists who had taken a more traditional symbolic approach, that is, representing cognitive processes by the operation of rules on symbols, attacked the Rumelhart and McClelland model on a variety of fronts. They criticized the connectionist program for failing to reproduce some of the learning patterns observed in children, as well as for errors in predicting the past tense of verbs and details of implementations, as well.

Two connectionists, Brian MacWhinney and Jared Leinbach of Carnegie Mellon University in Pittsburgh, responded to this challenge by developing a neural net that learned past tenses even better that the Rumelhart and McClelland model had done. Feeling a rush of pardonable pride in their creation, MacWhinney and Leinbach were emboldened to issue a challenge of their own: "If there were some other approach that provided an even more accurate characterization of the learning process, we might still be forced to reject the connectionist approach. . . ."

Many English verbs, the "regular" ones, form the past tense by adding "ed." Thus, I walk today while I walked yesterday. Other English verbs, the "irregulars," use a variety of endings to form the past tense:

swim/swam,
do/did,
see/saw,

and so on. Children display a classic learning curve when they absorb the intricacies of regular and irregular verbs. They learn slowly at first, then their competence drops markedly when they try to use the "ed" ending on all their verbs. After this stage, they seem to master both regular and irregular verbs at a steady rate until nearly full competence is reached.

The MacWhinney-Leinbach neural net had two intermediate layers of one hundred neurons each. They trained their network on 1650 verbs, 1532 regular and 118 irregular. After training the network for a day or two (an eternity in computation), they found synaptic weights that gave the net full competence except on 11 of the irregular verbs. MacWhinney and Leinbach claimed that if the back-propagation algorithm had been left to run for a few more days, their network would have reached full competence.

Such competence begged the question, however. When they tried the network on 13 irregular verbs that it had never encountered before, the network correctly predicted the past tense in only four cases. Since one test of any proposed model of how we think would be its ability to generalize from previous experience, the MacWhinney-Leinbach network clearly had a long way to go.

The challenge that MacWhinney and Leinbach so incautiously issued was met by Charles Ling at the University of Western Ontario in London, Canada. Ling developed a Symbolic Pattern Associator (or SPA) that outperformed the MacWhinney-Leinbach network in all respects, even its ability to mimic the learning curve of a child.

Theoretical psychologists and cognitive scientists have been using symbolic systems like Ling's for years. The symbolic approach to associating patterns uses simple rules such as the following:

$$XY\text{ing} \rightarrow XY\text{ung}$$

Such a rule would apply to the verb "swing," among others, where *X* is s and *Y* is w. The rules are added to "decision trees" that through repeated examples, are constructed by a special procedure used for most symbolic pattern association programs. When the training period is complete, programs like Ling's SPA may be let loose on new and unfamiliar verbs.

The SPA outperformed the MacWhinney-Leinbach network in every department. It had a higher score on the training set and a much higher score on previously unseen verbs. It took only seventy seconds of computer time to train to this level of competence; it mimicked a child's learning curve for past tenses much more closely than the MacWhinney-Leinbach network; and at the end of the day, it offered a set of rules that were far more transparent to the cognitive scientist than a table of thousands of meaningless numbers.

One example proves nothing, of course. Yet the way a rival system met the challenge of two experienced connectionists illustrates what may be a trend in the making. As we examine the real powers of neural nets, we will doubtless discover many areas of genuine competence for them. Yet, as many experiments have already shown, they are far from being miniature brains with universal competence. Their popularity will undoubtedly fade until they find their place as but one of many strategies for solving real problems.

Symbolic Pattern Associators, on the other hand, can hardly be expected to become popular. At least not until someone renames them "sym-brains," or something even more impressive.

Postscript

Throughout the neural net debacle, many computer scientists have kept a reserved silence on the subject. Those who knew the theories of complexity and computation, in particular, must have suspected that connectionists didn't. For the rest of us, neural nets have had a something-for-nothing quality, one that imparts a peculiar aura of laziness and a distinct lack of curiosity about just how good these computing systems are. No human hand (or mind) intervenes; solutions are found as if by magic; and no one, it seems, has learned anything.

6

Genie in a Jar

The "Discovery" of Cold Fusion

On March 23, 1989, chemists Martin Fleishmann and Stanley Pons stood the world of science on its head by announcing that they had achieved what hundreds of nuclear physicists had failed to achieve: sustainable nuclear fusion. The announcement could not have been more stunning if the pair had captured a genie in a jar. At a press conference hosted by the University of Utah, Pons's home institution, the two chemists described their apparatus. Not a billion-dollar hot-fusion reactor, but a simple glass electrolytic cell, some heavy water, a palladium electrode and a platinum one. The total cost of the apparatus came to perhaps a hundred dollars.

The fusion of two atoms into one releases tremendous amounts of energy, even more than the fission of a single atom into two. A hydrogen bomb, which operates by nuclear fusion, releases vastly more energy than an atomic bomb of the same mass. Although scientists developed working fission reactors before they constructed the first atomic bomb, things are the other way around with fusion. Several decades after the first successful hydrogen bomb test, scientists have yet to build a successful fusion reactor.

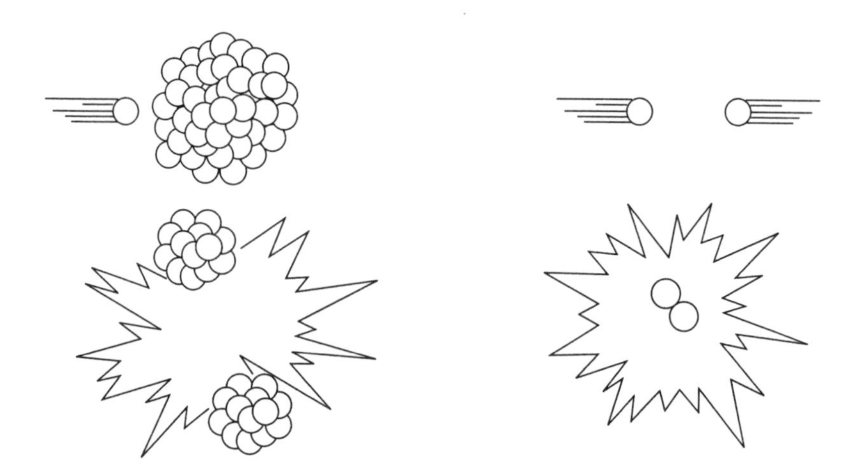

Fission and fusion.

To a world increasingly dependent on energy, a fusion reactor would prove an inestimable boon. Unlike fission, no exotic heavy metals like uranium are needed to drive the fusion process. It would work with simple atoms like hydrogen and fuse them into new atoms such as helium, essentially what happens in the center of the sun. Since almost all the energy that we enjoy at present comes directly or indirectly from the sun, one might say that we have been using a giant fusion reactor from the beginning of human life on Earth. The fuel for a fusion reactor closer to home is abundant, to say the least. Every molecule of water contains two hydrogen atoms.

Successful, sustainable fusion, although a real possibility, has a magical air about it. It would be a true genie, able to grant any and all energy wishes at the flick of a switch.

The process we might call "hot fusion" would work by driving atoms together under tremendous heat and pressure to ensure that fusion takes place. Current designs require not a jar but something like the enormous doughnut-shaped vessel called a tokamak. Inside the tokamak, a sophisticated plasma sustained by powerful magnetic fields sets up a frantic dance of atoms and particles that is statistically guaranteed to produce collisions between two nuclei in the plasma. The tremendous energies are necessary because atomic nuclei have similar charges and repel each other. Once atoms begin to fuse, however, the energy released keeps the process going and provides additional energy in superabundance. At the time of the earth-shaking announcement by Fleishmann and Pons in March 1989, nuclear physicists had coaxed the fusion genie into the heart of just a few reactors and for only the briefest appearances, milliseconds at a time.

Reactions to the announcement ranged from exaltation and greed to fear and scorn. Because Fleishmann and Pons were well-respected scientists, each with many publications to his name, the press felt safe in declaring the dawn of a new age of cheap and abundant energy. Average people, watching the news on their television sets, felt a positive surge of confidence in the future. Fortune 500 companies perked up their ears. Fusion in a jar? Time to get in on the ground floor. Within hours of the press conference and for days after, calls poured in to the University of Utah, many from firms offering to make the technology widely available—in return for a license!

Privately, more than a few nuclear physicists who assumed that the revelation had already been checked by their colleagues, felt simultaneously amazed and let down. Imagine! None of the high-tech equipment, none of the extensive theorizing, was necessary. A pair of smart-aleck chemists had beat them to the punch and made the whole hot-fusion effort look ridiculous. But as they learned more details of the Fleishmann–Pons experiments, they became increasingly skeptical. It was a bit like hearing that someone had made a nuclear reactor from a soup can. The skeptics had plenty of questions. For one thing, Fleishmann and Pons must have found lots of neutrons or protons coming from their jars. Did they? Had they found any of the typical fusion products such as helium or its isotopes?

More than a few chemists, aware that big-time physics had captured the lion's share of the headlines for years, may have smarted from time to time at what they took to be a certain air of superiority among physicists. Without a doubt they felt a certain glee at the thought of upstaging those fancy hot-fusion reactors with a simple electrolytic cell.

The actual announcement made by Fleishmann and Pons, even the paper they distributed at the conference, was maddeningly short on details. Within days of the announcement, hundreds, indeed thousands, of scientists around the world attempted to duplicate the experiment. Although a number of laboratories claimed to have duplicated Fleishmann and Pons's results, many more did not. Indeed, so many of the more careful attempts at duplication failed that the experiment came to be regarded as nonrepeatable, a cardinal sin on the face of things. The full story, however, has two parts. One is technical, the other human.

Fleishmann and Pons reported that their electrolytic cells produced more heat than they consumed. They also reported neutrons, but not at the correct energies and in numbers that were far below what physicists would expect if the pair had really captured the fusion genie in a jar. Explanations for the anomalous heat and the low neutron count have since appeared in the rather large literature on cold fusion. Is the dream dead? I'll come to that at the very end. In the meantime, the other half of the story reveals how Fleishmann and Pons came to believe that they had captured the fusion genie. They were already aware that others might not find the experiment repeatable. All too frequently, even their own cells failed to produce excess heat. Their reaction to this frustrating but apparently unavoidable feature of cold fusion illustrates what happens when more or less ordinary people get caught up in extraordinary events. Fleishmann and Pons lost their scientific patience. They gambled.

Dreaming of the Genie

It was Fleishmann who first dreamed of harnessing fusion in a jar. As an electrochemist, he was well aware how smallish electric currents passing through a chemical solution could sometimes cause reactions that would otherwise require exotic conditions. For example, to split sodium chloride (table salt) into its constituent atoms of sodium and chlorine by heat alone would take a temperature of 40,000 degrees centigrade. But if one merely dissolved the salt in ordinary water and passed a modest

4-volt current through it, the sodium and chlorine atoms would separate into ions, charged atoms with an excess or defect of one electron.

For years Fleishmann had been intrigued by the near magical power of the metal palladium. When used as an electrode, it could soak up hydrogen ions like a sponge. Could it be that if enough ions entered the palladium, they would be forced into fusing? In 1984 he had just taken early retirement from the University of Southampton in England. That year he visited his former student and erstwhile coresearcher, Pons, in Utah. The two had already published numerous papers together in other areas of chemistry, but this time, fusion was very much on their minds. Fleishmann wondered if it wasn't time to take a flyer on this wild idea. One night over a glass of whiskey in Pons's kitchen, Fleishmann said, "It's a billion-to-one chance. Shall we do it?" Pons replied, "Let's have a go."

The two had already done some library searching and turned up some intriguing hints that low-energy fusion might be possible, after all. Did they know that they were not the first to dream?

In 1926 two German scientists, Friedrich Paneth and Kurt Peters, had reported the spontaneous transformation of hydrogen into helium in the journal *Die Naturwissenschaften*. The announcement created quite a stir at the time, not because it meant free nuclear energy, but because it meant a new source of helium that could be used as a safe lifting gas in dirigibles and airships. Helium was extremely rare and difficult to produce at the time. The Paneth–Peters process involved the absorption of hydrogen gas by finely ground palladium. The two had measured small but quite definite amounts of helium apparently produced by the process. Later, however, they had to retract their claims when another researcher carefully repeated the experiment and found that the helium was being emitted by the experiment's glassware!

In 1927, Swedish scientist John Tandberg attempted to adapt the Paneth–Peters process to electrolysis. Ordinary water served as a source for hydrogen ions. The electric current in the electrolysis vessel forced the hydrogen into a palladium electrode where Tandberg thought it would be catalyzed by the metal, producing in the process "helium and useful reaction energy." In a patent application he claimed to have found a "significant increase in efficiency," meaning that the reaction seemed to consume less energy than it should, a hint that some energy was actually being produced. Tandberg's application was refused on the grounds that his description of the process was too sketchy.

Perhaps because of the first successful H-bomb tests, the 1950s witnessed an upsurge of interest in fusion. In parallel to the hot-fusion research effort then getting underway, people still sought an easy route to limitless energy. In 1951, Argentinian dictator Juan Perón announced that a secret national research facility, directed by former Nazi scientist Ronald Richter, had successfully tested a fusion reactor. When the effort was discovered to be bogus, Richter was arrested and the three hundred scientists working on the project were sent home. The "reactor" consisted of a giant chamber in which a spark gap was supposed to ignite a fusion reaction in a mixture of hydrogen and lithium gas.

Apart from soaking hydrogen or deuterium into metals, there turned out to be another route to low-energy fusion. In 1947 British physicist Charles Frank discovered a new particle which he called the mu-meson, or muon for short. The new particle had the same charge as an electron, but 207 times the mass. Among other things, this meant that a muonic hydrogen atom (in which the heavy muon replaced the light electron in its orbit around a single proton) would have $\frac{1}{207}$ of the diameter of an ordinary hydrogen atom. Such atoms could be brought much closer together before they experienced the repulsion that keeps nuclei from approaching each other too closely. Physicists theorized that such atoms might fuse more readily. In 1956 Louis W. Alvarez at the University of California at Berkeley actually observed muons catalyzing the fusion of deuterium atoms. The finding was genuine but the method did not lead to a fusion reactor; the cost of producing muons was (and remains) too great.

The work of Frank and Alvarez began a stream of legitimate research that continues to this day. Among the more recent researchers in the field was Steven Jones, a physicist who himself would become the catalyst in a more human reaction. Working at Brigham Young University, a mere 50 miles south of the University of Utah, Jones had already been studying muon catalyzed cold fusion but had recently shifted his research effort under the urging of a colleague, Paul Palmer. In 1986 Jones and Palmer began experimenting with electrolytic cells using ordinary water at first, then 10 percent heavy water and various other mixes. Jones thought he detected excess neutrons emanating from the cell but his neutron detector was too crude to make the measurement precise enough. He spent the next year and a half building a sophisticated neutron spectrometer, and by 1988 he felt ready to confirm his earlier finding of a slight excess of neutrons (above the background level) coming from his electrolytic cells. But back to our main subjects.

When Fleishmann and Pons began their first cold-fusion experiments in 1984, they agreed to work in secret. Part of the reason for this was undoubtedly the potential embarrassment of a colleague discovering their oddball project. Moreover, to conduct the experiment on an official basis, especially if they were to seek funding, they would have to comply with troublesome university procedures designed to protect scientists, subjects, and the world at large from various hazards, including nuclear ones. In deciding to work in this manner, the two began a pattern that would continue right up to the time of the fateful press announcement—and beyond. An obsession with secrecy, especially later when they thought they were witnessing fusion on a regular basis, would undermine their work in critical ways.

After some initial experiments in Pons's home, the two moved their venue to a laboratory in the basement of the University of Utah chemistry building. Their first run of serious experiments was designed as a probe into the possibilities. They began with simple electrolytic cells and used lithium deuteroxide as an electrolyte. This was a compound of lithium (the next heavier element to helium), deuterium, and oxygen. They used a small block of palladium as the cathode, of course. They had what they thought were good theoretical reasons for believing that fusion would take place in the palladium. Using a formula called the Nernst equation, they calculated that when the palladium became fully charged with deuterium, the metal would exert an enormous pressure on the particles, some 1,027 atmospheres, well in excess of the pressures achievable in today's hot-fusion reactors. Naturally, this excited the pair and heightened their expectations further. As it would later turn out, they had misapplied the formula. The actual pressure predicted by the formula was well below the one required for fusion.

Fleishmann and Pons knew that if fusion took place in their cells, neutrons might emanate from the palladium. They installed a simple neutron detector of the type used as a safety monitor near atomic reactors. As electrochemists, they were more interested in heat than radiation, of course. They immersed their cells in a calorimeter, essentially a water bath kept at a constant temperature. The temperature difference between the cell and the bath would give them an idea of how much heat was escaping from the cell. Using a standard formula, they could convert the rate of heat leaving the cell into watts, a measure of the power contained in the heat. The current and voltage entering the cell, when multiplied together, also gave them a figure in watts. They only had to compare the two numbers to know whether a cell was actually producing more power than it took in. Because the experiment was not an official one, they paid for most of the materials and equipment themselves.

As they ran their experiments, they discovered to their amazement and great delight that sometimes their apparatus produced excessive heat, more than they could account for on the basis of the charging current that flowed through the cells from their power supply. It also exceeded the amount of heat that could be caused by any known chemical reaction. The phenomenon was elusive and only appeared after the palladium cathodes had been charging for several days or more. They began to refer to the cathodes that seemed to produce excess heat as "live," the other cathodes as "dead." As experienced scientists, Fleishmann and Pons were aware that some subtle effect of the electrolysis or some flaw in their measurement might just be responsible for the seeming phenomenon. During the early period, they proceeded slowly, as if reluctant to be drawn into the clutches of a false hope.

But something strange happened in the basement laboratory one night while the experiments ran untended: One of the cells exploded, seriously damaging the floor and some of the nearby equipment. Fleishmann was away in England at the time, and the degree to which this event excited the two can scarcely be imagined. Pons phoned Fleishmann directly. Fleishmann's more or less immediate reaction was terse: "We had better not talk about this on the phone." Had the genie paid them a visit?

This event became famous after the fateful press conference. Cold-fusion supporters cited it as evidence of a nuclear explosion. Cold-fusion skeptics thought that the most likely explanation involved a natural cavity inside the palladium. A buildup of deuterium in the cavity, even at the relatively low pressure dictated by the Nernst equation, might well be enough to explode the block when the deuterium concentration reached the strain limit of the metal. Fleishmann and Pons decided to scale the experiments down, substituting a cylindrical palladium foil for the earlier palladium blocks. A new excitement gripped the pair, and they began to work in earnest.

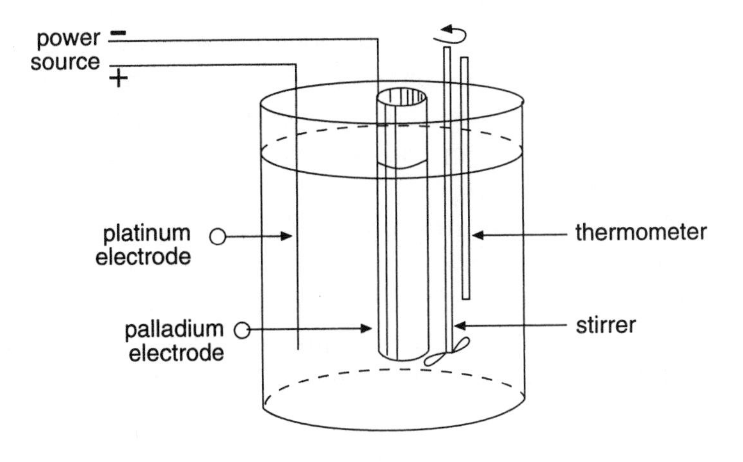

The new cold-fusion cell.

At this stage, their typical experiment emerged. They would allow a cell to run for several days to "charge up" the palladium with deuterium, then begin careful temperature measurements to see if their cell was generating heat. On several occasions cells appeared to produce excess heat, typically between 10 percent and 25 percent. Using a process he called "scaling up," Fleishmann theorized that based on these figures, a much larger cell with a larger palladium electrode would produce 4 watts of heat for every watt of electrical power input to the cell. The two would cite these figures at the March 23 press conference as if they had actually achieved them.

The simple neutron detector also sometimes showed a neutron count that seemed to exceed the background level (as measured by the same instrument). On one occasion, the count went 50 percent above background level. The two must have been aware, even vaguely, that if fusion was occurring in their cells, the amount of heat being given off should have been accompanied by a lot more neutrons than they were detecting.

The fisherman who snags an old boot experiences great excitement. It may not be fighting much, but if it's heavy, he might well think, "I have something!" Fleishmann and Pons had measured heat that was well in excess of any chemical reaction they could think of. Theorizing vaguely as they went, they simply postulated an "unknown nuclear process," one that would produce relatively few neutrons, even none at all. Because their experiments sometimes took weeks, involving long waits while the palladium electrodes charged up, the two continued to pursue other scientific and professional activities. Life went on as before but each man carried a terrible secret in his heart. They had something.

In 1988 Fleishmann and Pons decided to take the next step. It was time to design a series of more sophisticated experiments in which they could simultaneously vary a great many parameters such as electrolyte, electrode shape, current, and so on. In this way they hoped not only to eliminate the ephemeral, on-again-off-again nature of their results, but also to determine some of the design constraints on a useful cold-fusion reactor. To vary so many parameters at the same time would require a lot of cells and a great deal of sophisticated equipment, not to mention a new neutron counter. The two decided to draw up a grant proposal that they subsequently submitted to the Basic Energy Program of the U.S. Department of Energy (DOE).

Had things proceeded normally at this point, Fleishmann and Pons would have received the DOE grant in due course and done the new suite of experiments. They might well have discovered that no combination of parameters would result in a reliably reproducible experiment. They might even have taken a nuclear physicist into their confidence. Of course, it is very hard to take anyone into your confidence when you think you're sitting on the most important scientific discovery ever made. In fact, it must have taken the pair a long time to word the proposal so as to describe what they were up to without explaining the application they had in mind.

It takes nothing more than normal amounts of ego to explain what happened next. If you think you're sitting on the most important scientific discovery of all time, you live in fear that some other researcher may beat you to it. Enter Steven Earl Jones at nearby Brigham Young University, the scientist just down the road who had recently built a new neutron spectrometer in his pursuit of low-energy fusion.

It may sound coincidental that the DOE sent Fleishmann and Pons's grant proposal to Jones for refereeing, but the selection was natural. Jones was well known for his work in low-energy fusion, had a good research track record, and had evaluated other proposals for the DOE. When Jones read the proposal, he boggled. Two chemists a mere 50 miles away were proposing to carry out experiments that were alarmingly similar to his. Jones took the unprecedented step of asking the DOE funding director if any objections might be raised to his contacting the applicants. Jones's desire to contact the pair seems to have been motivated by a generous spirit. Perhaps they would like to use his new neutron spectrometer. They might consider working with him or, at least, coordinating publications. Jones, in any event, did not think he was sitting on the discovery of the millennium.

When Jones contacted Fleishmann and Pons in the fall of 1988, the pace of events picked up considerably. Jones, who planned to address a meeting of the American Physical Society in May 1989, submitted an abstract early in the new year. Their hands now forced, Fleishmann and Pons visited Jones in his Brigham Young laboratory. After discussing their results, the three agreed to submit separate manuscripts, simultaneously, to the prestigious science journal *Nature* on March 24.

The effect of the meeting on all three participants can only be guessed. Clearly, the two sets of experiments only served to reinforce the impression of each scientist that he was on the right track. At the same time, the agreement provided fertile ground for suspicion. What if the other party reneged and published first? For example, did Fleishmann and Pons violate the spirit (if not the letter) of this agreement by sending a paper on cold fusion well before March 24 to the *Journal of Electroanalytical Chemistry* without telling Jones? Enter the university administrators.

When Pons first approached President Peterson of the University of Utah to announce that he and Fleishmann had apparently discovered a process that produced fusion at room temperature, great excitement spread through the upper echelons. If cold fusion were a reality and the claims of Fleishmann and Pons were correct, the university would become immensely famous and wealthy. But in sharing the dreams of wealth and fame with the pair, the university became vulnerable to the same fear of being scooped. The air of secrecy spread to the administrative offices.

For one thing, University of Utah administrators and legal staff worried about Steven Jones at nearby Brigham Young University. Almost from the beginning of contact between their own two chemists and the physicist at Brigham Young, there had been a parallel contact between the two universities. However, University of Utah administrators were far more excited about the financial prospects than their counterparts at Brigham Young. Jones claimed no excess heat from his own experiments, merely a low level of neutron emission that barely exceeded background levels. On the other hand, if there was credit to be shared, the Brigham Young people were not prepared to take a back seat to anybody.

Even as University of Utah attorneys began to file patent applications, the two sets of administrators met to clear the air and come to an agreement about proceeding jointly. On March 6, the presidents of both universities, along with Jones, Fleishmann, and Pons met at Brigham Young to discuss cooperating in the matter of publication. All agreed that on March 24 the two research groups would each submit a paper to *Nature,* sending them off in the same courier package.

Shortly after, something spooked the University of Utah. Was it the fear that the press had already heard rumors of the cold-fusion work? Would Jones's work undercut their patent claims? The university abruptly decided to hold a press conference on March 23, one day before the agreed-upon date for filing the two papers.

The crisis was coming to a head. In deciding to hold the press conference, they did not even inform their own physics department! Jones, who learned of the conference only a day in advance, felt deep disappointment. As far as he was concerned, the March 6 agreement ruled out such an announcement. Pons and Fleishmann did not feel that way, nor did University of Utah administrators. In fact, shortly after the press conference, a reporter asked one of the Utah principals at the March 6 meeting whether he knew of any similar work elsewhere. The reply was negative.

Up to the moment of the press conference, Fleishmann had both good news and bad news. On the one hand, the DOE had approved their grant application for some $322,000 and the *Journal of Electroanalytical Chemistry* had accepted their paper, the one that would confirm their discovery and square things with the world of science. After all, in the normal course of events, scientists must publish first, then await review of their work by peers. The bad news was the neutron data. They needed confirmation of neutrons, but did not want it from Jones since it would mean that Jones would share the credit. Fleishmann contacted friends at Harwell, the British atomic research establishment, to see whether they could duplicate his setup and measure neutrons. Although Harwell could not comply in time, the prestigious British nuclear laboratory began intensive secret experiments that would last until June of that year. In the meantime, Fleishmann and Pons hired a radiologist to take gamma-ray measurements in the vicinity of heat-generating cells.

Fleishmann and Pons felt they needed another eighteen months of quiet research, but events seemed increasingly out of their hands. The stakes in their gamble had suddenly doubled. It was one thing to bet a few years of part-time research on what Fleishmann had called a "billion-to-one" chance. As long as no one knew, the damage to reputations would be minimal if it all came to nothing. But now they had a lot to lose. They evidently decided to make the best of a bad situation and go for broke. If they had really captured the fusion genie in a jar, all they had to do was announce their findings and other scientists, especially physicists, would merely confirm the finding. Theirs would be the glory, for ever and ever.

The Nightmare

It began with lights, cameras, and more action than either Pons or Fleishmann had ever imagined. The press conference at the University of Utah on March 23, 1989, brought representatives of every major network and wire service, along with reporters from major newspapers and magazines. To this assemblage Fleishmann and Pons announced that they had achieved sustained fusion in a jar. In particular, they implied that their experiment was very easy to replicate and that there would be no particular difficulty in scaling it up to useful reactor size. Headlines around the world trumpeted the dawn of a new age of cheap and limitless energy. Amid the blaze of lights and the incessant questions, Fleishmann and Pons must have felt like "the thermodynamic duo," as some reporters called them. They had become major celebrities overnight.

The heady times extended throughout the evening, on into the next day, and for weeks afterward. Meanwhile, scientists around the world were attempting to find out more about the experiment. Requests for information poured into the University of Utah by mail, e-mail, and telephone. What were the dimensions of the cell? What current and voltage were appropriate? What type of palladium had the pair used? The lucky ones got preprints of the paper that would later appear in the *Journal of Electroanalytical Chemistry.* Copies of the precious preprint, incomplete and riddled with errors as it was, multiplied like lemmings.

The press conference had blindsided normal scientific routine. Attempts to replicate the experiment were often based on incomplete, inaccurate, or misleading descriptions of the experiment. Scientists could only forge ahead with what they had, filling in the details with guesses about what Fleishmann and Pons had done. This created another potential abuse of the scientific method: If a scientist failed to get the proper experimental results, Fleishmann, Pons, and their rapidly growing body of supporters could simply claim that the scientist in question hadn't used the right equipment or procedure.

When major laboratories announced failures to duplicate the experiment, Fleishmann and Pons may have been annoyed, but the thermodynamic duo had only themselves to blame. They had told the scientific world that cold fusion was a reality. The attempt by a group at the Massachusetts Institute of Technology illustrates the difficulties. Led by physicist Stanley Lockhardt, a team at the M.I.T. Plasma Fusion Center attempted to duplicate the experiment. They didn't want to be accused of using the wrong materials or methods, and the preprint was inadequate for their purpose. The group was forced into the rather unscientific procedure of using network videos of the Utah laboratory to see how many cells Fleishmann and Pons had used, how the cells were wired, and so on. They had to obtain the diagram of a typical cell from a copy of *the Financial Times* of London!

The M.I.T. group tried for a week, once they had set up their apparatus, but got none of the effects that Fleishmann and Pons had claimed for their process. They reported "no fusion" to the press.

The nightmare was just beginning for Fleishmann and Pons. But for a while the dreamlike atmosphere persisted. On April 10, the *Wall Street Journal* printed the headline COLD FUSION

EXPERIMENT IS REPORTEDLY DUPLICATED. At a Dallas press conference, a team of scientists from Texas A&M University reported 90 percent excess heat output from one of their experimental cells. Fleishmann and Pons were tremendously pleased and excited. As if to heighten the euphoria, another team at Georgia Tech announced neutrons emanating from their fusion cells. In the next four days more confirmations and partial confirmations came from a few other North American labs, as well as from India and Russia. Then, on April 15, two graduate students at the University of Washington in Seattle, Van Eden and Wei Liu, announced that they had measured large amounts of tritium (a possible fusion byproduct), emitted by their experimental cell. They too held a press conference. As one columnist put it, science was now being done by press conference.

The Fleishmann–Pons strategy of going for broke seemed to be paying off, and Fleishmann's doubts evaporated. Cold fusion was undoubtedly real.

But the nightmare returned almost immediately. On April 15, Georgia Tech retracted its claim to have found neutrons. Their counter, it turned out, was sensitive to heat. Then nine days later, even as more confirmations came in, the Texas A&M group also retracted their claim. They had not grounded their thermometer properly, and the current flowing through it had heated it up, giving an exaggerated reading.

At this point, the number of labs reporting negative results began to exceed the number that claimed positive ones. Fleishmann and Pons began speaking less and less to the press. They apparently granted interviews only to journalists who appeared to believe in cold fusion. On May 18 *Nature* published a damaging article by Richard D. Petrasso of M.I.T. The physicist demolished the claims of Fleishmann and Pons to have measured neutrons via gamma rays at the 2.224 MeV energy. Worse yet, on May 25 the University of Washington graduate students Eden and Lui reported that they had made a mistake: They had not found tritium after all, but another tri-atomic molecule that was fairly common in their environment.

Although some laboratories and individual scientists continued to report some neutronic or heat effects, the major laboratories now began to sound the death knell of cold fusion. On June 15 the British laboratory at Harwell, which had finally completed its million-dollar intensive cold-fusion experiments (under Fleishmann's original direction), reported no fusion. Over the summer, the California Institute of Technology, Oak Ridge National Laboratory, and other major national labs reported no fusion. A special panel convened by President Bush under the aegis of the Department of Energy spent the summer visiting various sites and examining firsthand the evidence for cold fusion. In the fall, this panel also reported that there was no evidence for cold fusion as claimed by Fleishmann and Pons.

By August 7, when the University of Utah sponsored the first annual meeting of the National Cold Fusion Institute, it must have seemed strange to the two electrochemists to find themselves surrounded by two hundred "believers," as they were then coming to be called. From this point on, however, the number of believers began to dwindle steadily. The dream was dying, and somewhere (but not in a jar) the fusion genie laughed a hollow laugh.

On October 23 no one could find Pons, and his house had been put up for sale. On January 1, 1991, the University of Utah announced the resignation of Pons. Peterson had already resigned as president of the University of Utah that summer. In the next summer, on June 30, the National Cold Fusion Institute closed permanently.

As individuals, Fleishmann and Pons had taken an extraordinary gamble and lost. They lost, in the end, not because they were anxious for Nobel prizes or great wealth. Such factors may explain why they gambled. But they lost because they were wrong. They were wrong about the neutrons and wrong, apparently, about the excess heat being radiated by their electrolytic cells.

Yes, We Have No Neutrons

Perhaps the most damning evidence against cold fusion came from the living presence of Fleishmann and Pons themselves at the press conference. They claimed not only heat but also neutrons were given off by the fusion process. They attributed this neutron radiation to a nuclear process in the palladium as it absorbed its limit of deuterium ions. Nuclear physicists who watched the press conference and heard the figure "watts of heat" would have known that the number of neutrons that should have accompanied such heat would have led to quite serious health problems for the pair long before the press conference. Instead of less than a hundred neutrons per second, the pair should have been bathed in a deadly spray of some thousand billion neutrons per second.

Another criticism that physicists leveled at the pair concerned their original measurement of neutrons by their crude counter, called a BF3. Experiments with this counter at Georgia Tech revealed that it was very sensitive to heat. When brought near even a moderate heat source, it would begin to register a higher count of "neutrons." The Georgia Tech team that had earlier announced a replication had discovered the fatal flaw in the BF3 counter by experimenting with it.

Throughout the neutron debate, as the evidence for neutrons began to crumble before their very eyes, Fleishmann and Pons stuck to their theory that an "aneutronic" nuclear process was responsible for the heat. In the long run, the neutrons didn't really matter.

Turning Up the Heat

At the press conference, Fleishmann had claimed "4 watts of heat output for every watt input." Later, on May 8, Fleishmann would turn up the heat by announcing experiments where their cells put out fifty times the energy put in. Perhaps with the neutron side of things going so badly, it was time to emphasize the heat even more. But the heat they had measured came under attack almost as swiftly as the missing neutrons.

When Fleishmann spoke at Harwell five days after the fateful press conference, a scientist in the audience asked him whether he and Pons had done any control experiments. Had they, in fact, done identical experiments with ordinary water replacing the heavy water in the cells? Fleishmann replied, "I'm not prepared to answer." What reason could he have had to reply in that manner? In a normal scientific gathering, it would sound simply childish, like a small boy saying, "Won't tell you." Some controversy came to surround this question. Had they or hadn't they run the control? At the Harwell meeting, Fleishmann might have hoped that his audience would think that some vital but secret patent claim hinged on the control experiments, at which they could all at least nod their heads and murmur sagely. Later, when someone asked Pons the same question, he replied that control experiments with light water were "not necessarily a good baseline."

If it were the case that Fleishmann and Pons actually had run experiments with ordinary water and observed results similar to those with heavy water, it would throw further doubt on their findings since they had implied that the heavy water was an essential ingredient. Could the calorimetry of two experienced calorimetrists be off, or did they have something? As far as physicists were concerned, it was not fusion.

We may never know for sure just what caused some of Fleishmann and Pons's cells to overheat. Overheating has certainly been detected (and explained) in other cells, however.

As far as Frank Close, a physicist at Oak Ridge National Laboratory, is concerned, Fleishmann and Pons may have fouled up their calorimetry, nevertheless. They used a type of calorimeter called "open." In other words, the gases that evolve at the electrodes may escape into the laboratory atmosphere. Such gases would not be counted in the heat budget. Leaving the water (heavy or otherwise), the individual deuterium and oxygen atoms would be free to recombine. Essentially, the hydrogen or deuterium "burns" in the presence of oxygen, and some of the heat created may reenter the cell by radiation. This may well be the source of "excess heat" that Fleishmann, Pons, and many of their believers found. It is true, nevertheless, that many subsequent attempts to detect the anomalous heat with closed calorimeters have failed.

Nathan Lewis, an electrochemist at the California Institute of Technology, was another scientist who was instrumental in driving nails into the cold-fusion coffin. He found, among other things, that electrolytic solutions of the type used by Fleishmann and Pons would not register the correct heat if they were not vigorously stirred.

Believers continue to believe in the face of a world seemingly grown hostile to the possibility of cold fusion. If some scientists are angry with Fleishmann and Pons, however, they can be forgiven more readily than the thermodynamic duo can. For the better part of a year, the world of science had been turned upside down. In the normal order of things, a scientist will (1) have an idea or insight, (2) design a series of experiments to test the idea, (3) publish the results if the experiments establish something new, and (4) await the results of others who attempt to duplicate the experiment. If the result is new, the scientist gets the credit.

The furor initiated by Fleishmann and Pons amounted to a sophisticated guessing game. They never completed step 2, and in the absence of a reasonably complete description of their experiment, step 3 was essentially omitted. Moreover, the urgency attached to such an important claim converted the media and the Internet into temporary journals, in effect, a role to which they are particularly unsuited given the time it normally takes to replicate an experiment. In the absence of step 3, however, perhaps the massive exchange of information via the electronic media prevented step 4 from taking forever.

The cold-fusion fiasco nevertheless illustrates the interplay of theory and experiment under extraordinary conditions. When confronted by the absence of neutrons from their electrolytic cells, they posited an "aneutronic process." This sounds very scientific, but it only means "a process that doesn't emit neutrons." Fleishmann, Pons, and their supporters would stand their ground, committed to saving the cold-fusion hypothesis by invoking new and fabulous theories, thought up on the spur of the moment. To John Huizinga, a physicist at M.I.T. and chair of a Department of Energy panel to examine cold fusion, the new claims were reminiscent of Langmuir's Laws of Bad Science (see the end of chapter 1), particularly numbers four and five:

4. Fantastic theories contrary to experience are suggested.
5. Criticisms are met by ad hoc excuses thought up on the spur of the moment.

There was no shortage of believer-theorists to create theories that would account for the anomalous, irreproducible nature of the Fleishmann–Pons reaction. If there was going to be a paradigm shift, they would get the credit for the theoretical breakthrough. At the same time, skeptic-theorists tried to account for the unexpected evolution of heat. Suppose that at least some of the measurements were real. How to account for them without assuming that a fusion genie inhabited the lucky cells? Variations of the following theory also made the rounds.

In soaking up the deuterium ions, some obscure electrochemical process also stores up energy. Later, when the palladium electrode nears saturation, this energy is released in large amounts. In other words, for the long hours that many calorimeters appear to be losing heat, they are actually storing it up as potential energy, like winding a clock. They might store the potential energy in a rearrangement of the lattice atoms and their deuterium visitors. Then the lattice begins to fall into a new arrangement in which a lot of electrons and nuclei run "downhill" in an atomic sense, settling into new, lower-energy configurations. The lost energy appears as heat.

The point remains that, strictly speaking, no one can say for sure that something strange didn't happen in at least some of the cells. Believers, including Fleishmann and Pons at latest count, cling to the ever-diminishing probability that the fusion genie actually visits electrolytic cells.

Postscript

The gradual die-off of public interest in cold fusion was predicted by Langmuir's Laws of Bad Science; but cold fusion is not quite dead, even for Fleishmann and Pons. Early in the 1990s, Eiji Toyota, president of Toyota, Inc., took an interest in the possibility of cold fusion and decided to try a long shot of his own. He funded a complete cold-fusion laboratory for the dynamic duo near the French city of Nice.

At last report, Fleishmann and Pons were happily running new versions of their experiment and building a commercial-scale experimental cold-fusion reactor. Whatever we may think of the science, how can we not wish them luck?

7

Biosphere 2 Springs a Leak

On September 2, 1991, eight people entered a sealed environment called Biosphere 2 for a two-year stay that was intended to make scientific history. Set in the Arizona desert with the Catalina Mountains looming in the background, Biosphere 2 seemed a distinctly alien structure, as if some Martians had settled in for a stay. Or was it the other way around?

For most people, the story of Biosphere 2 began a few months before its official sealing. That's when the media began to report that eight brave people called "bionauts" were about to enter a new world without leaving Earth. Television screens and newspaper photos displayed a huge, futuristic steel-and-glass structure. Inside the domes and arched glass halls were a rain forest, an ocean, a desert, a marsh, and grasslands, five "biomes" in all. It was the Earth in miniature, all but polar environments finding a home under the glass.

The Biospherans stood for numerous photo-ops, smiling and keen in their bright red, futuristic jumpsuits. The scene must have reminded more than a few viewers of the movie *Close Encounters,* in which Earthlings selected by aliens prepared to board a huge UFO for parts unknown.

Jubilant Biospherans about to be sealed in.
(AP/Wide World Photos. Used with permission.)

The media read a short list of the bionauts' credentials and mentioned some of the scientists behind the project, including Carl Hodges, director of the Environmental Research Lab at the University of Arizona. Hodges, who also directed research and development on Biosphere 2's "agriculture biome" had earlier told *Time* magazine, "This is not an academic exercise meant to generate Ph.D.'s." That was about as far as Hodges could go publicly. The statement seemed to carry an implicit warning that the project would not have any academic goals. Just what goals would it have?

The purpose of Biosphere 2, as explained in the media, seemed to change with every story. To some of the Biospherans, the impressive structure would make unique ecological experiments possible. It was a "living laboratory." Mark Nelson, a project director, expressed the goal more expansively. Biosphere 2 was a "cyclotron of the life sciences." In it they would pursue a brave new science called "biospherics."

Project Scientific Director Tony Burgess went even further, calling Biosphere 2 "the Chartres cathedral of the Gaia Hypothesis." By this he meant, possibly, a place where one could worship the goddess Gaia. This was the name bestowed on the Earth as a whole (considered as a living organism) by the prominent biologist Lynn Margulis.

Other stated goals, as gathered by various media, had a more sober flavor. The sealing in of the Biospherans was itself an experiment to discover, as the prestigious *Economist* put it, "whether man can design and live in a self-supporting biosphere in which the environment provides everything for life." The *New Republic* took a more businesslike tack, claiming that the purpose of the project was "to develop the technology necessary to colonize other planets with biosphere structures."

It was perhaps not difficult to reconcile these goals. Who could say that a prototype space colony could not also be a cyclotron of the life sciences, not to mention a kind of cathedral? But the credulity of many was strained to learn that Biosphere 2 may have had yet another mission. It was owned and operated by a private company called Space Biosphere Ventures (SBV) Inc, and one reporter claimed that SBV hoped to sell its ecotechnology to NASA and to the European Space Agency. It was Space Biosphere Ventures that had dubbed the structure "Biosphere 2" in reference to that other biosphere in which we all live—the Earth, or Biosphere 1.

Yet another goal emerged during the period of intense media attention prior to closure: SBV had set up an extensive tourist facility adjacent to the biosphere site and it also planned a huge theme park that would co-occupy the 2,500-acre Sunspace Ranch. After all, a protospace colony that operated a biological cyclotron and developed new space technology might well benefit from the addition of a theme park.

In the view of a few independently minded reporters, Biosphere 2 itself was already something of a theme park. A certain atmosphere of unreality at the site disturbed Jeanne Marie Laskas and Peter Menzel, two reporters who probed Biosphere 2 in the summer before closure. In a story that appeared in the

August 1991 issue of *Life,* Laskas and Menzel puzzled over just what the mission of Biosphere 2 was. They succeeded in getting only the most convoluted answers from staff and management.

"There are no tidy answers to that question. Ask the Biosphere 2 project leaders what they're trying to prove and you are in for some pretty scrambled reasoning. These people go off on tangents—long involved tangents. All of a sudden they'll be talking about when Kennedy was shot, or Planned Parenthood." Laskas and Menzel were further troubled by the science-fiction atmosphere that seemed to pervade the project. "Here people talk very casually about spending their retirement years on Mars . . . as if it were France." What about the proud new science of Biospherics? They asked Arthur W. Galston, a Yale botanist; his reply was "garbage." In particular, Galston leveled one main criticism at the project: "They're not asking a question; they're saying, 'Suppose we build this thing. What's going to happen?' That's a question, in a way, but it's not science."

If there was one element in the Biosphere story that caught media attention more than any other, it was the idea of complete and absolute separation of Biosphere 2 from its parent planet. In short, the bionauts might as well be going to Mars! If the seal were ever broken, woe betide them. The ark would sink in a sea of scientific and public ridicule.

The Biospherans settled into their new environment with high hopes and the high morale that comes with positive publicity. The story in *Life* was just one bad wave in a sea of otherwise good news. *Discover* magazine gushed that Biosphere 2 was "the most exciting scientific project to be undertaken in the United States since President Kennedy launched us toward the moon." The *New York Times,* the *Boston Globe,* and other major newspapers caught the fever and also published favorable stories. For network news, the video copy couldn't have been better. Eight brightly clad, mostly young Biospherans were going somewhere new and mysterious. They were completely sealed off from the rest of the world for two years!

What then, was Biosphere 2 all about? The confusion of purpose revealed at every turn barely hinted at another, darker mission of which all the avowed public purposes turned out to be mere facets. This mission involved an ark, Biosphere 2, and a quasireligious group that made a covenant to build a new humanity that would migrate to the stars. Behind SBV, we find a fascinating cast of characters acting out a fantasy that came slowly to fruition over twenty-five years. Before turning to the all-too-human side of the Biosphere 2 story, I will set the scientific and technological stage for the disaster in the desert.

Science or Technology?

If there was any science behind Biosphere 2, it was surely embodied in the question stated by the *Economist:* "Whether man can design and live in a self-supporting biosphere in which the environment provides everything for life." But as the botanist Galston had put it, the question was not a scientific one. It would involve testing the hypothesis you'd get by removing the "whether" from the previous quote. But such a hypothesis would not be scientific because it involved no generality. If Biosphere 2 succeeded, the answer would be yes, but what else would we know? In fact, the great experiment of Biosphere 1 had already answered the question: Yes.

But as a technological question, applied to the amazing steel-and-glass structure in the Arizona desert, a yes answer might be worth a lot of money to Space Biosphere Ventures. The company apparently expected to live or die in the marketplace by the success of its $60 million demonstration: It is possible to seal up a representative sample of the Earth's living matter with people, add sunlight, and expect the whole closed ecosystem to survive, in some robust sense, for several years.

As I will demonstrate later, the group knew of some experiments in this area but badly misinterpreted the results. In the meantime, SBV had confused science with technology. Biosphere 2 was not a scientific project but a huge technological gamble, one that failed in the first instance because it ignored so much science, especially biology and ecology.

Despite its scientific shortcomings, Biosphere 2 is nevertheless an impressive structure with a lot of impressive things inside it. Located on the 2,500-acre Sunspace Ranch 35 miles north of Tucson, the T-shaped structure took six years to plan and build. Its three sealed-in acres are portioned into seven special areas. Five natural "biomes" span the bar of the T: a rain forest, a tropical savanna, a desert, an ocean, and a marsh. Two other areas, which according to SBV represent the human biomes called city and agriculture, occupy the stem of the T and actually consist of a garden area where the Biospherans grow much of their food and the Biospherans' living quarters and nerve center, a miniature city called Micropolis.

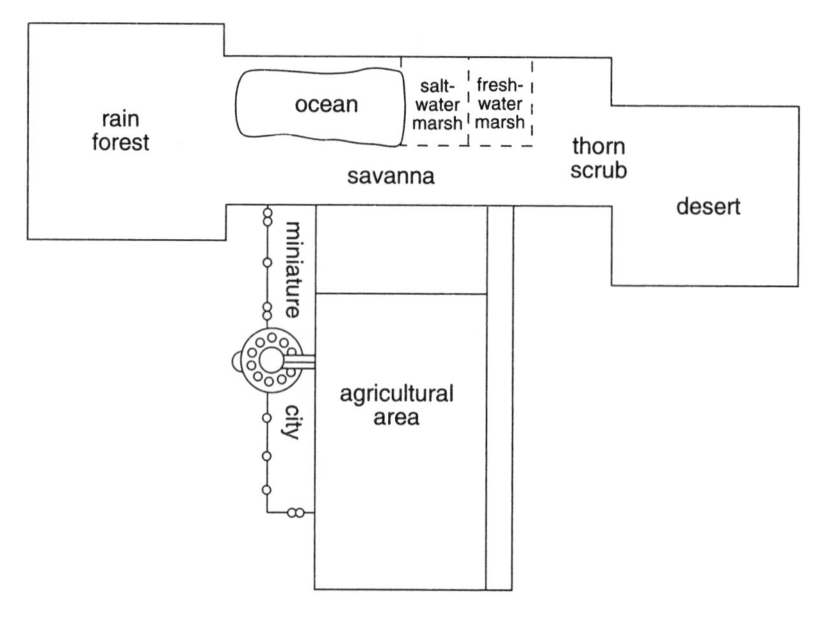

Layout of Biosphere 2.

One gets the strong impression, on hearing of these areas, that someone in SBV was dreaming metaphorically rather than thinking objectively. The entire complex, it turns out, is organized like a human body. It has "lungs," a large circular room where air pressure is adjusted; "kidneys" where water is purified by passing it through algae; a "nerve center" where computers and other equipment monitor conditions; even a spinal cord of bunched cables that carry signals from one end of Biosphere 2 to the other.

The designers of the various biomes that SBV installed in Biosphere 2 were all experts, some more and some less, in their fields. The desert, designed by botanist Tony Burgess, was modeled after the "fog deserts" of Baja California. Although fog deserts occupy a few humid spots around the world's coastlines, they hardly typify what most people think of as desert, a biome that would have been impossible to include owing to the high humidity of the enclosure. The fog desert in Biosphere 2 nevertheless included many species of cacti and other succulent plants imported from Baja, along with a few tortoises and other desert reptiles.

A transition zone of thorn scrub from Madagascar and Mexico led into the savanna designed by Peter Warshall, an anthropologist. The savanna contained plants from Africa, South America, and for good measure, Australia, some forty-five species of grasses alone. There were small animals here, including some toads, lizards, and the occasional tortoise that wandered in from the desert.

A stream that ran through the savanna entered the adjacent marsh designed by Walter Adey, an experienced display designer at the Smithsonian Museum in Washington, D.C. At Adey's direction, SBV imported great numbers of plants and animals from the Florida Everglades, including fish and frogs. Adey, who had been looking for an opportunity to build a marsh, also agreed to build the Biosphere 2 ocean. His marsh drains into an ocean, which he wanted to make much larger because he feared that the planned size would generate too little oxygen. But SBV balked at the cost, an extra $10 million. The ocean had its own coral reef and supported, at least when first installed, some one thousand species of plants and animals.

Ghillean Prance, Director of the Royal Botanic Gardens in Kew, England, designed the rain forest that lay on the other side of the ocean. He hoped it would be just large enough to be viable. Grafted onto a miniature mountain of artificial rock, the rain forest looked slightly miserable, like a garden in early spring. It would take years before some of the trees reached even moderate height. In the meantime, bats and bushbabies lurked in its foliage while hummingbirds flitted among branches in search of pollen-laden flowers. Here also, a colony of termites aided in the decomposition of "dying material." A miniature cloud hovered uncertainly about the summit of the artificial mountain in the middle of the rain forest. Not the result of some marvelous ecological interaction, the cloud issued from a mist machine that recycled water from the Biosphere basement in the form of a fine mist to keep the rain forest humid.

To fit five biomes under one roof, SBV personnel and their consultants had to make some disappointing compromises. For one thing, they could not fully represent each Earthly biome from which they extracted species for Biosphere 2. In each case, the team responsible for a given biome had to make hard decisions about what to take into the ark and what to leave behind. The marsh team, for example, imported no alligators or water birds, key predators in the Everglades ecosystem. The relatively small size of the 3-acre structure left little enough room for creatures that like to range. Alligators, for example, might well wanter into Micropolis looking for a good meal. Water birds would break their necks on the low grass ceiling above the marsh.

Down on the "farm," the Biospherans would grow their own food on a half-acre area. In recycled earth and organic waste they would grow corn, tomatoes, soybeans, and other crops. Each Biospheran would spend an allotted four-hour period each day working in the garden. This included not only planting, cultivating, and harvesting the crops, but gathering eggs from the chickens, milk from the goats, and meat every now and then from the Vietnamese potbellied pigs. There were rice paddies, as well, where the edible African Tilapias fish would, in turn, eat algae and water ferns, fertilize water with their waste, and thereby help the rice to grow. Spiders, wasps, ladybugs, and lacewings were added to the mix to keep destructive insects away from crops. Whenever the rain forest produced a banana or a papaya, the Biospherans could supplement their diet with fresh fruit.

Back in Micropolis, each Biospheran had his or her own 360-square-foot apartment but shared common dining and recreational facilities. In the city's nerve center, the Biospherans could communicate with the outside world via telephone, fax, or through a TV linkup. Here also, computers were linked both to the outside world and to a myriad of sensors spread throughout the complex of pipes, pumps, valves, and control systems. All the air in the building had to be continuously recycled and monitored for its mix of oxygen, nitrogen, carbon dioxide. Although no electric power was permitted to enter the building, a 5.2-megawatt solar power plant apparently provided enough electricity to run the entire complex.

Finally, the seal. As SBV already knew, and as any engineer faced with the problem of building an airtight structure will tell you, the job is impossible. There will always be tiny leaks. It was the contractor's job to minimize leaks, however, by employing a special seal that attached glass-and-plastic panels directly to a space frame of steel members. As a design goal, SBV adopted a leak rate of one percent of the atmosphere per year as the maximum acceptable. But even this modest-sounding goal may have been unrealistic. With over 50 miles of glazing and 12 miles of welded seams, it would take only a single hole one-tenth of an inch in diameter to produce a one percent leak. The real trouble would be the microholes, those one-thousandth of an inch across. Impossible to detect, they could easily be present in great enough numbers to leak one percent or more. SBV could do little more than hope.

Springing Leaks

Within just a few weeks of closure, trouble came to the Biospherans in the form of an innocent accident. While operating a thresher machine in the agricultural biome, Biospheran Jane Poynter accidentally cut off the end of her finger. Needing medical attention, Poynter was permitted to leave through the biosphere airlock. She returned a day or two later, her finger in a bandage, but no worse for wear. None of this, even had the media known of it at the time, would have caused any comment had Poynter returned as she had left. True, she had broken the seal, but who would blame her, considering the seriousness of the accident? But Poynter had also brought a duffel bag with her back into Biosphere 2. Rumors of the incident began to circulate, rumors alleging that Poynter had returned with fresh food and a new set of seals. The latter would be used to "prove," according to the same rumors, that Biosphere 2 had never been unsealed. SBV denied the rumors until January of the new year, when they finally announced that the accident had indeed taken place but that Ms. Poynter had returned to the biosphere with only a few irrelevant items like plastic bags, books, computer parts, and film in the mysterious duffel bag.

But distant rumblings of the coming media storm that followed had already sounded when Marc Cooper, a reporter for the *Village Voice* of New York, began to publish his investigations of Biosphere 2 and the people behind it. In April 1991 and again in July, two months before the Biospherans were sealed into Biosphere 2, Cooper's first exposés of the project appeared. He claimed that the Biosphere 2 project had been masterminded by a "new age cult" headed by John Allen, a poet and Harvard MBA. Cooper produced evidence from several sources that Allen intended to found a new humanity that would colonize Mars, leaving a decaying Western civilization to silently rot away.

Cooper was not alone. Others had picked up the trail even earlier, including reporter Victor Dricks of the *Phoenix Gazette* and the Canadian Broadcasting Corporation (CBC). In fact, SBV had threatened the CBC with a lawsuit if it dared to distribute tapes of its 1989 program in the United States. The CBC meekly complied.

In a second story, published in November 1991, Cooper struck again, this time with an exposé that a carbon dioxide scrubber had been installed just before closure. Shortly after this, project programmer Rocky Stewart resigned in protest over the CO_2 scrubber. He maintained, further, that SBV management had been misleading the public about several aspects of the Biosphere 2 project. As if to confirm Stewart's contentions, SBV pumped in 600,000 cubic feet of outside air into Biosphere 2 in order to maintain a failing atmospheric pressure. When Biospherans Linda Leigh and Roy Walford realized what had happened, they threatened to leave the project unless SBV made a public announcement. One by one, other media began to pick up the new, negative theme, and SBV found the atmosphere of Biosphere 1 growing cloudy.

Meanwhile from inside Biosphere 2, Biospheran naturalist Linda Leigh published a series of reports in *Buzzworm*, an environmental magazine. The articles demonstrated her touching determination to make a go of it:

> For me, Biosphere 2 has been an experience of the world of life unlike any other. The invisible world of atmosphere has had to coalesce with the visible world of plants, animals, soils, and rocks. My experience as a naturalist is different from any I have had previously, now concentrating intensively on one small area. I visit my study sites daily, sometimes hourly; I do not need to travel longer than three minutes to arrive in the rain forest, savanna, desert, ocean, or marsh biomes. My source of food is just downstairs. The wild areas are equally as accessible as my apartment, computer, telephones, and videotapes.

Negative publicity resulting from the Poynter incident and the CO_2 scrubber revelation, along with the continuing stories from Cooper and Dricks, finally spurred Ed Bass, funder of the project, to set up a scientific review panel in the spring of 1992 to recommend changes in the way that science would be conducted in Biosphere 2. But even as the panel studied the problems of scientific management and reporting, a new problem had developed in Biosphere 2, one that would administer the coup de grâce to the grand experiment's credibility. Here is the report by Linda Leigh filed early in 1993, nearly sixteen months after closure:

> By January 13 . . . the oxygen in our atmosphere had decreased from 21 percent, which is normal for the Earth, to 14 percent. I had mixed feelings about it. From one point of view it was a welcome surprise that would

> give us the opportunity to study processes of oxygen cycling in Biosphere 2, some of which may be the same as those occurring in Biosphere 1 (Earth). Alternatively, there was concern that we would need to add oxygen from outside during this closure if the animals inside, including humans, showed serious symptoms of oxygen deprivation.

Later, this was precisely the decision taken by SBV, to add oxygen. Management closed off the west "lung" and injected outside atmosphere enriched to a 26 percent oxygen content. This meant breaking the seal once again. Finally, the Biospherans could troop off to the lung to breathe.

> Seven of us stood by the opening of the west lung for the rush of oxygen that would come out when the door was opened. . . . I got a sudden impulse to run around the lung for no conscious reason, just an impulse which drove my legs. Odd because I'm not normally prone to run. But when I returned to my starting position, I observed that I was not panting, that in fact I didn't feel at all like gasping for air, whereas just fifteen minutes earlier I would have been terribly winded just walking up ten feet of stairs slowly. I was floored by this discovery. I felt like a born-again breather.

Even with breathable air back in Biosphere 2, the Biospherans experienced other problems—food for one. The agricultural biome had been designed by Roy Walford, in-house doctor and inventor of a low-protein diet. The Biospherans, who frequently went hungry, lost an average of 13 percent of their body weight and often felt on edge and temperamental as a result of continuing hunger. They became obsessed with all aspects of food, from watching it grow, to harvesting, cooking, and eating it. Tempers occasionally flared when something went wrong with the food supply, gathering, or cooking process. There was a near mutiny, for example, when peanut rations had to be imposed.

In the summer of 1992, the panel convened by Bass had filed a report recommending changes to the project. Chief among these were that the scientific credentials of those conducting experiments be upgraded to the Ph.D. level and that SBV end its proprietory attitude to experimental data. Science, claimed the panel, depends on sharing. Very little changed, however, in the way SBV conducted the "Mars colony" business, and by the end of April 1993 the entire scientific panel, chaired by Thomas Lovejoy of the Smithsonian Institution, resigned. Citing continuing difficulties with the SBV management, Lovejoy finally concluded in frustration, "The Biospherans will soldier on, but their two-year experiment in self-sufficiency is starting to look less like science and more like a $150 million stunt."

The bionauts emerged on September 26, 1993, to a lukewarm world, at best. The scandals and revelations about the real purpose behind Biosphere 2 had cooled media interest considerably. Worse news now followed. Many species had gone "locally extinct" as biologists say.

The hummingbirds and bees had vanished, as did some finches, ocean fishes, and plants. Estimates of total species losses ranged between 15 and 30 percent. At the same time, some species had gotten out of hand. Shrubs and grasses had begun to overrun the desert, mites ate the potatoes and white beans, while the Vietnamese potbellied pigs ran amok in the vegetation and had, in turn, to be eaten by the Biospherans. Trigger fish ate too many of their reef colleagues and had to be removed. Worst of all, cockroaches had multiplied explosively over the two-year period and invaded all the terrestrial biomes as if they were so many New York City apartments.

The most interesting thing to happen in Biosphere 2 from an ecological point of view was the loss of so much oxygen. Outside scientists soon pinpointed one of the causes: The soil of Biosphere 2 had been enriched with extra fertilizer, and this had encouraged the growth of more than the normal amount of bacterial decomposers. Many of the bacteria, it turned out, were aerobic, meaning that they respired, or used oxygen.

Despite these problems and despite its obvious failure to operate a closed ecosystem for two years, SBV dubbed the experiment a success. This meant, evidently, that because the failure in the oxygen supply had tempted scientists to speculate about the cause, humanity had learned something as a result.

Synergia Ranch

If Biosphere 2 had a starting point, it was at a mysterious place in Texas known as Synergia Ranch in the late 1960s. The times were heady for some. The countercultural revolution, signaled by the appearance of hippies, granola, and acid rock (among other things), set the stage for new cultural ideas. It encouraged a new political awareness that fueled the anti-Vietnam war movement, a new spiritual awareness that led some to dabble in Eastern religions, and an environmental awareness that led others to experiment with self-sufficient rural communes. The three themes seemed to have merged at Synergia Ranch.

Named after R. Buckminster Fuller's concept of synergy, Synergia Ranch was home to what reporter Marc Cooper called a "new age cult" that was run by John Allen, a poet, engineer, and self-styled bioscientist. Synergia Ranch first surfaced in a book called *The Commune Experience* by Lawrence Veysey, a University of California historical researcher. The book revealed details that were later confirmed by Kathelin Hoffman, a former member and one-time confidant of Allen. As leader or guru, Allen is reported to have run his commune not as an open, cooperative society but as a tightly controlled group to which he, Allen, preached a message of societal decline brought on by the "sleep" into which humanity at large had fallen. Allen further taught that only those who could "wake up" would deserve to escape the decay. Only those who could awaken to their full human potential deserved a place among the stars.

Allen's teachings seem inspired by the writings of the little-known G. I. Gurgieff and his followers. Gurgieff, an Armenian-Georgian who flourished as a mystic teacher during the first three decades of this century, cast an enormous spell over followers during this period. He encouraged "inner work" by which disciples could bring about the desired state of "awakening" and "self-remembering" that opened the door to all higher things. He stressed absolute obedience and trust to his disciples.

Members of the group at Synergia Ranch used false names, perhaps to foreshadow their coming personal transformations. Allen, who called himself Johnny Dolphin, exercised the iron control required to bring his disciples to the required state of awakening. He gave them tasks that kept them fully occupied on the inner work, subjected them to long mealtime monologues, and according to a number of ex-members, used both psychological and physical coercion, in the form of public dressing-downs and beatings, to impose his will. He had been known to deliver fierce lectures to individuals who strayed from the true path of "harmonious" development, as Gurgieff had called it. In retrospect, many of these incidents appear to have had personal motives, as when Allen particularly favored a new female inductee or felt his own position of control and authority being threatened by another male.

Allen urged his group to engage in craftsmanship, no doubt impressed by Gurgieff's stress on the practice. On many weekends, visitors to Synergia Ranch might pause to buy some earthenware pots or woven goods. One such visitor was to have a profound effect on the group.

In 1974 a young man named Ed Bass visited Synergia to buy some handmade furniture and stayed to talk philosophy. As everyone quickly learned, he was the second son of Texas Oil multibillionaire Perry Richardson Bass and a billionaire in his own right. Suddenly, for Allen, the sky was the limit—or no longer the limit. The group's new human ingredient put a whole range of new projects in reach, and it did not take Allen long to think up one worthy of their newfound financial clout. Humanity, being largely asleep, would simply stumble blindly around, messing up over and over again. The world was going to hell in a handbasket, and only those worthy of being saved would survive. But where would the chosen ones live? Where else? Mars.

Group members would not only awaken into a superior humanity, they would engineer their own escape from a decaying world. No doubt, if Allen could demonstrate the technology that made survival on Mars possible, and if he could convince space agencies to adopt the technology, thousands would flock to him, seeking to be part of the new humanity. He would, in the process, found a new synthesis of science, technology, the arts, and personal development. Biosphere 2 was born.

It was all there in a book of John Allen's poetry, available in the Biosphere 2 gift shop.

Lebensraum

As a child,
On a board with clay
I marshalled Alexandrian empires;

In my teens,
Roaming the western states
I was passion's Khan;

Now,
suave with galaxies,
From this cell I plot escape.

After visiting Biosphere 2 halfway through the first experiment, journalist Michael O'Keeffe of *Buzzworm* magazine wondered in print, "Does Biosphere 2's mission have anything to do with saving this planet? Or is it about ditching the Earth for Mars?"

If the ark called Biosphere 2 really did begin at Synergia Ranch, it explains much about the seeming confusion of purpose and the strange atmosphere that some reporters would later discover at the Biosphere 2 site. In accord with the real purpose of Biosphere 2, it would be essential to (1) develop the ark technology, (2) retain control of the technology in order to retain control of the ark, (3) impress humanity with its leadership in science and technology, and (4) develop a facility for direct recruitment of new followers. It must do these things, moreover, without a hint of spiritual idealism. What better disguise than a corporate front?

Under this analysis, SBV's main goal would be to demonstrate the viability of the Biosphere 2 technology. It would keep certain aspects of that technology secret, however. This explains the stage-managing of the media and SBV's secretive attitude toward sharing experimental data and technology, not to mention its paranoia about bad publicity and even its forbidding tourists to take pictures on the Biosphere 2 site. The new science of Biospherics, meanwhile, would realize goal 3 and perhaps the theme park would act as a recruitment point in accord with goal 4.

But the hidden agenda of Biosphere 2 does not fully explain the bad science that permeated almost every aspect of the project. For that we must hypothesize two things: a willingness to be blinded by a fantasy and ignorance of the relevant science.

First of all, Allen and his cohorts clearly assumed that by throwing five "biomes" into a pot called Biosphere 2, the whole system would automatically do in miniature what the Earth did as a whole. It would "pop," as marine aquarium enthusiasts call it when the flora and fauna within their tanks arrive at a self-sustaining equilibrium. They believed this, in part because certain experiments in closed ecosystems seemed to point toward an almost indefinite sustainability and robustness in such systems. As we shall see presently, Allen gravely misinterpreted the results of this study, absorbing only its positive implications and completely ignoring the warnings of its authors.

Given such misinterpretations, it still remains unclear how anyone could actually believe that five biomes could be crowded into such a small space and continue in their original form. Problems of scale pervade the project.

Ecologists still don't know the minimum size that a biome can be and still survive. They do know, however, that if it shrinks (usually through human exploitation) below a certain extent, it is practically doomed. The same thing is true of pockets of species. If the range or breeding population of a particular species becomes too restricted, it too is doomed. Natural fluctuations in weather, food supply, and habitat can reduce already small numbers very easily to the point where a single reproductive failure causes what ecologists call a local extinction.

In the rain forest, a phenomenon known as the edge effect nibbles away at any abrupt margin (usually of human creation) by encouraging a succession of plants and animals that belong to grassland, scrub forest, or other biomes to invade the forest. At the same time, the lack of a canopy completely alters the habitat below the large trees at the edge. Rain-forest biodiversity drops within a hundred or so meters in this zone, even as new plants and animals take over. Left to itself in the middle of an open Amazonian field, the Biosphere 2 "rain forest" would largely disappear within a few decades.

Although few people are aware of the fact, some 80 percent of the world's oxygen comes not from trees but from oceanic algae, individual and colonial cells equipped with chlorophyll. Covering some 75 percent of the Earth's surface, our oceans may be barely adequate for the job. Yet the miniature "ocean" was expected to function in relation to Biosphere 2 as did the Earth's oceans in relation to Biosphere 1. Walter Adey, who helped design the Biosphere 2 ocean warned SBV that it was not nearly big enough to trap sunlight sufficient to support the air-breathing life forms under glass. His warning was disregarded by Allen and SBV management. Ultimately, Adey was pushed out of the project. David Stumpf, the University of Arizona researcher who helped design the agricultural biome, says that he warned Allen as early as 1986 that the large number of animals in the Biosphere might well produce elevated CO_2 levels. Given that the Biosphere 2 plans were already set, the only solution Stumpf could suggest was to install a carbon dioxide scrubber, a mechanical device that removes carbon dioxide from the air pumped through it. Stumpf recalls that this suggestion disturbed Allen deeply. Biosphere 2 would be a self-sustaining environment and that's all there was to it. Stumpf was literally shouted down by Allen and Margaret Augustine, Chief Executive Officer of SBV and longtime Allen crony. Stumpf later said, "What really disturbed me was how immature they were about it, their inability to be wrong unless they said they were wrong. . . . As soon as you deviate from what's drilled into their heads, they're lost."

Oxygen and carbon dioxide, two of the most precious gases in Biosphere 1, were clearly destined for a major imbalance in Biosphere 2 long before the first closure. Only the installation of the CO_2 scrubber (finally adopted by SBV) kept the carbon dioxide from skyrocketing. Unfortunately, this didn't prevent the oxygen from plummeting. To understand why, there is no alternative to looking under the hood of ecosystems generally.

Ecological Systems

The field of ecology, essentially a discipline within biology, began in the late nineteenth century when a few biologists began to recognize that nature was more than the sum of its parts. One could study separate plants and animals and yet still miss something vital. More than any single factor, the evolutionary theories of Charles Darwin and Alfred Russel Wallace had made it plain that the interactions of species, both with each other and with the environment as a whole, constituted a separate subject of more than theoretical importance.

In spite of the slow and steady progress of ecology, its classification of relationships and its many field studies of plant and animal assemblies, our growing concern with pollution and loss of species brought new urgency to the subject.

During the same period, a much smaller number of scientists sought to tease out the exact relationships that comprised ecosystems. They built "closed systems," sealed containers in which just a few organisms might seek a balance that promised to last a long time. Much of this research was funded by navies and by space agencies, both in the United States and abroad. The applications to humans spending a long time in a confined space seemed obvious.

What does it take to produce a closed ecosystem that is capable of supporting human beings, or for that matter, any air-breathing animal? The fundamental cycle involves two processes, photosynthesis and respiration. In photosynthesis the substance called chlorophyll captures light energy, transforms it into chemical energy, and makes it available for the creation of compounds like sugars, starches, cellulose, and other complex organic compounds. Many organisms use chlorophyll for just these purposes, including plants and algae. In the process, such organisms absorb carbon dioxide (CO_2) and water (H_2O) from the environment. Their photosynthetic apparatus splits the water molecule into free oxygen (O_2) and uses the hydrogen, along with the oxygen and carbon from CO_2, to synthesize compounds like glucose, a simple sugar. In the process, they store the energy of the sunlight in the chemical bonds of the glucose molecule. Biochemists write the basic formula in the following standard form:

$$6H_2O + 6CO_2 \rightarrow C_6H_{12}O_6 + 6O_2$$

Here, six molecules of water and six molecules of carbon dioxide combine to produce one molecule of glucose and six molecules of oxygen. For every atom on the left-hand side of the equation, there is a corresponding one on the right.

The reaction is certainly not as simple as the equation seems to imply, however. Photosynthesis involves many stages during which energy-carrying molecules are created and destroyed, protons are transported across miniature membranes, and glucose is built up into more elaborate sugars and starches, enabling the construction of proteins.

The other key reaction in living matter, respiration, runs in the reverse direction. Simple sugars are broken back down into carbon dioxide and water. Not everyone appreciates the simple but beautiful symmetry of these twin processes of life. Just as the reaction above characterizes the fundamental process by which plants get energy from sunlight, the reverse reaction characterizes the equally fundamental process by which animals use plant substances to develop usable energy. In the process called respiration, animals breathe oxygen to combine it with simple sugars like glucose to produce carbon dioxide and water. Their energy comes from the bonds of the glucose molecule:

$$C_6H_{12}O_6 + 6O_2 \rightarrow 6H_2O + 6CO_2$$

The two reactions between them comprise a fundamental cycle without which no sealed ecosystem worthy of the name could survive for long. To incorporate this cycle would require, at a minimum, the enclosure of a photosynthetic organism with an air-breathing one. A supply of light would power the photosynthetic reaction, and the food energy and oxygen thus provided by the photosynthesizing organism would sustain the second reaction in the respiring organism. To close the loop, the respiring organism would provide the water and carbon dioxide needed by the first organism. It may sound like an easy cycle to implement, but few biologists have felt tempted to try.

Since the 1950s, a small cadre of scientists have worked in what might be called closed ecosystems research. Generally underfunded, the field has barely progressed beyond its infancy. Typical of the early experiments were the ones carried out by R. O. Bowman and F. W. Thomae at Chance Vought Aircraft, Inc. As early as 1960, Bowman and Thomae had sealed a container of the alga known generically as Chlorella (the photosynthesizer) with one or two mice (the respirers). They ran eight trials, five of which had to be terminated owing to an excessive buildup of CO_2 or drop in O_2, within the sealed chamber. Bowman and Thomae had terminated the five trials for purely mechanical reasons when an equipment breakdown forced them to stop.

In each case, the system seemed to develop a certain tenuous stability. The most successful run lasted twenty-eight days. In it, a 4-liter culture of Chlorella supported the respiration of a single 30-gram mouse. During this period, oxygen content of the chamber air increased from 21 percent to a maximum of 30 percent. But a follow-up experiment, with two mice under the same conditions, failed. Here, the oxygen content varied between 20 and 31 percent until, after two weeks, it dwindled mysteriously to a mere 10 percent. Bowman and Thomae halted the experiment to relieve the oxygen distress of the two animals. Far from respiring, the mice were expiring.

What had caused the mysterious drop in oxygen? It didn't take long to find out once the two scientists analyzed the entire system. The mice feces, accidentally wetted in the chamber, had become the home to a new and unintended culture of aerobic bacteria. Animals are not the only organisms to respire with oxygen. While plants and algae may photosynthesize in the light, for example, they must respire at other times, using their own stored-up supply of food. Moreover, some bacteria occupy the same respiratory niche as animals in being aerobic, requiring oxygen to survive and multiply. This experiment had special relevance to the failure of Biosphere 2, as I noted earlier.

But even the successes showed enough variability to put the whole concept of a closed system into question. More complex systems that included algae, a rat, and a fungus proved more difficult to manage. One question that dogged researchers posed both philosophical and practical problems: What was a closed system, anyway? Frieda B. Taub, a biologist at the University of Washington in Seattle, has followed the closed-system experiments from the beginning. In her opinion, if one takes the idea literally, nothing should go in or out for the duration of the experiment. She would allow light to enter a sealed chamber, for example, but not the lab chow that fed the rats and mice in early experiments. The resulting chamber would be called materially closed but energetically open.

The champion materially closed systems were developed by Clair E. Folsome, formerly of the University of Hawaii, and Joe A. Hanson, of the Jet Propulsion Laboratory in Pasadena, California. During the 1970s and early 1980s, the two scientists demonstrated that materially closed ecosystems are possible, but not without their dangers to the organisms involved.

For example, Folsome and Hanson reported a series of experiments involving sealed flasks in 1986. In one of these experiments, the scientists set up a large number of 2-liter flasks in which they placed varying combinations of a wild microbial mix with varying numbers of the crustacean *Halocaridina rubra,* a small marine shrimp that is barely over a centimeter long. The wild mix, in this case, consisted of some Hawaiian beach sand. Virtually any sample of damp sand, mud, or soil taken almost anywhere in the world will be found to contain vast numbers of protozoa, algae, and bacteria. The samples taken by Folsome and Hanson were no exception. Their microinhabitants found themselves sealed up in the flasks for 441 days to work out a balance with the tiny shrimp and each other.

Folsome and Hanson exposed their flasks to systematically varied amounts of light on a twelve-hour cycle. At the end of 441 days, all components of the miniature ecosystems seemed to be alive and well, albeit in some of the flasks the shrimp population had declined. However, all declines observed by the scientists reached some kind of equilibrium within about sixty days of closure and thereafter maintained small but robust numbers.

Neither Folsome nor Hanson were much surprised by the results. Folsome, for example, had already kept one purely microbial system going for more than eighteen years, while Hanson had maintained a closed microbial/shrimp system going for more than five. Although the shrimp appear to be capable of indefinitely sustained existence under these conditions (reproducing as they go), neither scientist was sanguine about the possibility of larger closed ecosystems without a great deal of further work. They noted ominously in their 1986 paper, for example, that "attempts to establish systems containing other, larger metazoans (including herbivorous fish) thus far have all resulted in death of the metazoans within three months of closure."

In particular, Folsome and Hanson pointed out that species loss must be regarded as part of a normal process by which a closed ecosystem seeks equilibrium of its parts. Some species of plants, animals or other organisms may no longer be present when the system is finally reopened.

John Allen and the others were well aware of the Folsome-Hanson experiments but chose to ignore the downside. Ironically, the not-so-closed ecosystem within Biosphere 2 may have been trying to reach a new equilibrium, one in which there would be no humans!

Postscript

In his struggle to turn Biosphere 2 into a successful research facility, Ed Bass sacked the entire management team, including Margaret Augustine and John Allen, in March 1994. He sought to turn the enterprise around by setting up new procedures, such as allowing research scientists to use Biosphere 2 for limited periods of time. In the same month a new team of six scientists and technicians entered Biosphere 2 for a ten-month period. They would operate more like the crew of a large research telescope, maintaining the instrument for visiting scientists, as well as running some experiments of their own.

Bass finally took himself out of the picture in January 1996 by placing the entire Biosphere 2 operation under the direction of a consortium led by the Lamont Doherty Earth Observatory of Columbia University. Biosphere 2 continues to struggle for a place in the scientific sun. Although its structure makes it impossible to replicate experiments in atmosphere/plant interaction, other experiments are surely possible. Perhaps Biosphere 2 will some day contribute to a more livable Earth. Who but space cadets would want to live on Mars, anyway?

8

For Whom the Bell Curves

The Racial Theories of J. Phillipe Rushton

The *Geraldo* talk show, not normally a platform for the debate of important scientific issues, aired on March 8, 1989, with a promise to explore the racial theories of Professor J. Phillipe Rushton, a prominent psychologist at the University of Western Ontario in Canada. Rushton had recently achieved media notoriety for his theory that in a variety of human characteristics such as brain size, IQ, and social restraint, the three races he had studied always come out in the same 1-2-3 pattern. Using Rushton's language,

Mongoloids > Caucasoids > Negroids

Mongoloids, he claimed, have bigger brains (and higher IQs and more social restraint) than Caucasoids who, in turn, have bigger brains (and the other things) than Negroids. Rushton also claimed to have discovered a variety of human characteristics, such as penis size, in which the opposite pattern prevailed:

Negroids > Caucasoids > Mongoloids

Explaining his reasons for agreeing to appear on the *Geraldo* show, Rushton said he merely wanted to clarify his views for the American public. He may have expected a forum on the scientific issues raised by his work, but that is not what he got. Before showtime, a production assistant had coached the audience to express their feelings honestly, coaching them, in effect, to express disapproval: "If you hear something here that you don't like, let us know by a boo, a hiss."

The host, Geraldo Rivera, who may have sensed a good opportunity to express some politically correct outrage of his own, displayed little patience with the scientific issues. Instead, he zeroed in on the mild-mannered, square-jawed professor's appearance: "I can't get over, as I look at you, that you have a resemblance to Clark Kent. Is there a master race behind those glasses?" Sitting in the audience was Charles King, founder of the Urban Crisis Center in Atlanta. "Sir, do you think you're superior?" he demanded of Rushton at one point. Rushton, who did not quite answer the question, calmly replied, "A racist categorizes every member of a race in the same manner and then mistreats them if they belong to that race."

Geraldo, who went on to discuss the penis size issue at length, so to speak, never looked back. Rushton must have been sighing inwardly by this point. It was proving impossible to get a fair hearing in any public forum.

Ever since delivering his fateful paper at the San Francisco meeting of the American Association for the Advancement of Science two months earlier on January 19, Rushton had been continuously in the limelight. The media, who usually have reporters at large scientific gatherings in case there is any important announcement, wasted no time putting the Rushton story on the wire. What ended up in the minds of the millions who saw or read the stories would be hard to characterize, but it may have gone something like this: Some psychology professor had scientifically demonstrated that Asiatics were smarter than whites and whites were smarter than blacks. But when it came to penis size and lack of sexual restraint, it was the other way around.

To some people the theory sounded reasonable, and to others it sounded either whacky or racist. But many people, blacks, whites, and Asiatics together, must have scratched their heads and privately wondered. Could the theory possibly be true?

The media, which have a major responsibility to seek out the facts behind news stories, even science stories, did what they usually do when time is short and deadline pressures loom. They dug up the names of whatever science consultants they could find in their files. Collectively, they blitzed North America's major universities, and ended with print, sound, and video bites that barely touched the real issues. Only occasionally did a scowling scientific face complain that Rushton had emphasized "between group differences" at the expense of "within group differences," but that was as close as the debate ever got to ground zero. There was no mention of the unsoundness of the IQ concept, nor any discussion of the other theories on which Rushton had leaned so heavily. Most of the discussion centered on Rushton's use of secondhand data, not to mention his analysis of it.

Some psychologists and other social scientists criticized Rushton's work, particularly his use of secondhand statistics and grand averages. Rushton, after all, had relied heavily on surveys undertaken by a wide variety of researchers to arrive at his conclusions. Some of them may have been a tad unreliable, but when you averaged them all together, *Voilà!* The interracial differences emerged.

Without necessarily realizing it, the critics who went after his statistical methods were tackling a tar baby. With every criticism, Rushton drew them in deeper. There was nothing wrong with using large-scale measurements. Not only were many of them reasonably accurate, but when you took a grand average, inaccuracies evened out. Secondhand statistics are not unknown in psychology, nor are grand averages. And for his other statistical methods, Rushton had merely followed standard practices of psychological research. The same methods, were they applied to a less controversial topic, would have passed muster in virtually any psychological journal. If continued, the debate might well have dragged arguers into the unwelcome conclusion that if Rushton's work was bad, the methods of psychological research were hardly better.

The real problems lay at a more fundamental level.

Shaky Foundations

Rushton's theory of racial differences consists of two major parts. First, there are the published measurements of body parts, IQ, and behavioral factors, along with Rushton's statistical analysis of them to establish averages and various correlations. Second, there are three theories developed by others that provide an interpretive framework for his analysis:

1. the theory of the intelligence quotient, or IQ, from his own field of psychology
2. the *r/K* theory from the field of ecology
3. the "out-of-Africa" theory from paleoanthropology

The following diagram shows the structure of Rushton's overall theory of racial differences.

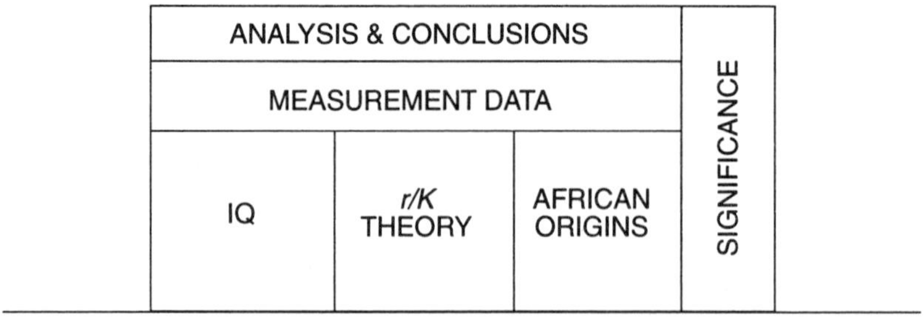

Rushton's theory.

He uses the first theory to interpret his IQ data. When it comes to "intelligence," after all,

Mongoloids > Caucasoids > Negroids

In the diagram, each component of the theory is symbolized by a stone block. While undoubtedly an oversimplification, the diagram displays those aspects of Rushton's theory addressed in this chapter. I have also added a small block to the right labeled significance, an issue that I will tackle last of all in a discussion of the bell curve itself.

The *r/K* theory of ecology asserts that the breeding capacity of animals is inversely proportional to the amount of time they spend nurturing their young. Rushton interprets "lack of social restraint" (as he has measured it) as well as penis size as indicators of high breeding capacity. When it comes to breeding capacity,

Negroids > Caucasoids > Mongoloids

Incidentally, the terms Negroid, Caucasoid, and Mongoloid are used routinely in the anthropological literature. They refer to the races that we know publicly as blacks, whites, and Asians. Whatever you may think of the scientific terms, they are at least stable (as scientific terms must be in order to be useful) and not subject to revisions every few years as the public terms seem to be.

The theory of two British paleoanthropologists opines that the human races originated directly or indirectly from Africa in three separate radiations. The Negroids were the first to develop, but they stayed in Africa. The Caucasoids were the next to develop, but they left Africa for western Asia. The Mongoloids were the last to develop. They colonized eastern Asia and in the process became what they are. Each succeeding race was more highly evolved than its predecessor. This theory supports Rushton's conclusion that when it comes to large cranial capacity (not to mention intelligence),

Mongoloids > Caucasoids > Negroids

As it turns out, all three theories, particularly IQ, have major problems connected with them.

The Intelligence Quotient

As chapter 2 made clear, the theory of the intelligence quotient fails to be science. Any conclusions based on it, such as blacks being "smarter" than whites . . . excuse me, whites being "smarter" than blacks, are automatically undermined as science.

Throughout the twentieth century, the notion of IQ as an innate quality, not to say a heritable one, has permeated our culture. Throughout their checkered history, IQ tests have been used not just by educators anxious to measure academic potential, but by those who would establish differences between races. The most notorious predecessor of Rushton in this regard was Arthur H. Jensen, a professor of educational psychology of the University of California at Berkeley. In an infamous 1968 article in the *Harvard Educational Review,* Jensen argued that differences in average IQ test scores between blacks and whites in the United States are due largely to genetic differences in intelligence. The article had the mild-mannered title, "How much can we boost IQ and scholastic achievement?"

But public reaction to the article, inflamed by a story in the *New York Times* about Jensen's work, was anything but mild-mannered. At one point Jensen had to confront an auditorium of jeering students, just as Rushton would some twenty years later. Jensen's article was widely interpreted as blaming the failure of compensatory educational programs for blacks on blacks themselves. It was black genes, in particular, and not the field of educational psychology, for example, that one should blame for the failure of the programs. The storm of controversy created by Jensen's ideas lasted well into the 1970s. Among all the criticisms of Jensen's theories, that of Richard C. Lewontin, a well-known geneticist, stood almost alone. The problem, according to Lewontin, was IQ itself.

In the more recent debate over Rushton's ideas, the notion that his IQ foundation stone might somehow be at fault simply did not appear during the media furor of early 1989. In February of that year, for example, a consortium of campus groups at Rushton's home institution, the University of Western Ontario in London, Canada, sponsored a debate between Rushton and David Suzuki, a geneticist turned science writer. While the audience jeered and chanted slogans, Rushton drew a tape measure from his pocket and put it around his head. He sounded plaintive as he suggested that all the audience needed to do to confirm the truth of his theories was to go around and measure people's heads.

Suzuki, for his part, expressed anger that such research was condoned at the University of Western Ontario. He condemned the university for not producing somebody to debate Rushton's views scientifically. In characterizing Rushton's views as "racist," Suzuki fell into the mild-mannered professor's trap. Rushton politely pointed out that Suzuki had expressed little more than "moral outrage" during the debate and that he, Suzuki, had ignored the many measurements that supported Rushton's contention. "That is not a scientific argument," Rushton said. He was right.

Besides failing to have a scientific foundation, one cannot escape the fact that IQ tests have been devised by whites and cannot help but favor, in subtle ways, white children and adults who take the tests. Very little cultural bias is needed to account for the small differences that develop between the IQ scores of blacks versus those of whites. In fact, what might be called "black IQ tests" have been devised. In these, blacks tend to outscore whites. Enough said.

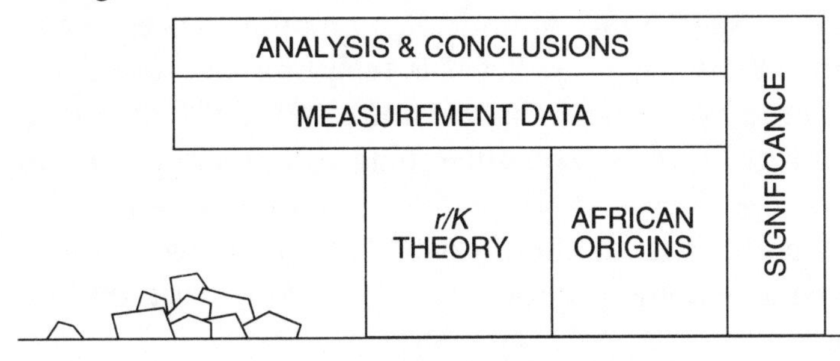

Rushton's theory.

The Theory of *r* and *K*

Chief among the behavioral characteristics in Rushton's racial theory are the linked qualities of social constraint, sexual constraint, and extended nurture of the young, with its associated smaller "litter" sizes.

Just as he uses the out-of-Africa theory to support his contention that brain size and IQ are both genetic traits that derive from more (or less) evolved ancestors, he uses the theory of *r*- and *K*-selection to support his contention that the more highly evolved races tend to show more social constraint, sexual constraint, and a greater concern and care for their young.

The theory of *r*- and *K*-selection developed first in the field of ecology, then spread to anthropology because it seemed to explain at a very coarse level at least, the differences between some animals. Animals such as frogs and mosquitos produce an enormous number of offspring and spend very little time in bringing them up. Other animals, such as horses and baboons, produce very few offspring, often only one, and invest a great deal of time in nursing them and training them, in effect, for their new life. The stakes are the same in both cases—survival of the species.

The letter *r* refers to the purely reproductive strategy of producing as many offspring as possible so that some of them, by pure chance perhaps, will survive. The letter *K* stands for the opposite quality wherein the parent or parents produce very small numbers of offspring, the better to nurture them to maturity.

Which strategy is better? Obviously neither. Each animal enjoys its own special mix of *r* and *K*, a mix that is presumably optimal for its own species. Yet it is tempting, because the *r*-strategy is used by so many "lower" animals and the *K*-strategy is used by "higher" ones, to assume that evolution moves in the direction of *K*-strategies. It is especially tempting to assume that humans, who have the highest *K* of all, are themselves more or less evolved depending on how much care they lavish on how few offspring.

The theory of r and K had its origin in the work of R. H. MacArthur and Edward O. Wilson in a classic 1967 paper on biogeography. In order to explain a well-known field observation that the reproductive rate of animals often goes up with increase in latitude, MacArthur and Wilson hypothesized that in the higher latitudes, the increasing likelihood of catastrophic weather events would severely reduce populations with little regard for their genetic endowment. In order to recover quickly from such catastrophes, the strategy of choice for a northern population is a high reproductive rate, r, even at the expense of certain other survival-enhancing strategies such as care of the young. In other words, a high value for r would mean increased fecundity, larger litters, earlier sexual maturity, and so on. Such traits, according to MacArthur and Wilson, would be called *r-selected.*

In kinder and gentler climes where the weather varies less violently, populations remain closer to the limit K imposed by available resources. In such regions survival traits that enhance the ability of individuals to exploit those resources more efficiently or to compete for them more effectively would be enhanced. Such traits are said to be *K-selected.*

Although "r- and K-selection occupy an important place in current thinking about life history patterns," ecologist Robert E. Ricklefs of the University of Pennsylvania notes that "a direct relationship between population growth rate or population fluctuations and life history characteristics has not been established."

By now, the alert reader will have noticed that MacArthur and Wilson attached high K-values to northern populations, not equatorial ones, precisely the opposite of what Rushton assumed!

Although ecologists spend no time worrying about human r and K, some anthropologists have at least considered the theory. But even they understand that it applies only when there are substantial differences between the organisms involved (like horses and frogs) and not present-day human groups. At best, one might suppose that a very early hominid had more offspring and spent less time rearing them than a modern human, but that would already be stretching a theory that in any case, has received very little support in the field.

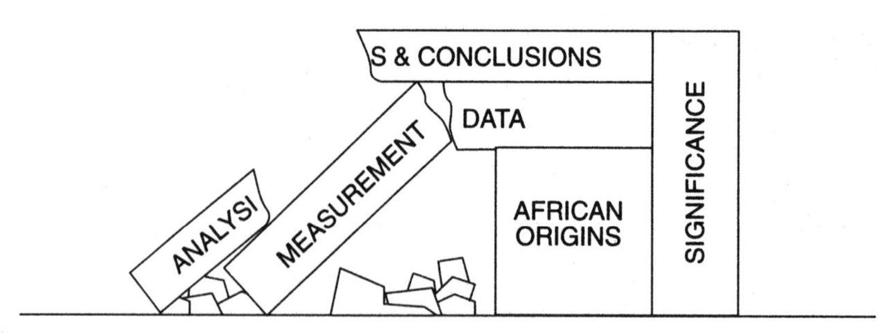

Rushton's theory.

Out of Africa

On the face of it, Rushton's theory concerns various human characteristics such as cranial capacity, measured IQ, nurturing behavior, sexual restraint, risk-seeking behavior, size of genitals, and about fifty other qualities altogether. His main papers, however, stress the characteristics just mentioned. After seeing a list of all the characteristics and the way that Rushton claims they manifest in the three racial groups, readers who recall the Darwinian connection might well ask, Does all this have anything to do with evolution?

As far as Rushton is concerned, it has everything to do with evolution, recent human origins, in particular. Rushton, who I suspect would rather be an anthropologist than a psychologist, has lately added a new wrinkle to the racial theories of Galton and Jensen. There is a reason, according to Rushton, that blacks, whites, and Asiatics show all these supposedly genetic differences. Enter the theory of Stringer and Andrews.

In 1988 Christopher Stringer and Peter Andrews, two paleontologists with the British Museum in London, published a very influential paper in the prestigious journal *Science*. Their paper, based on recent DNA studies both of apes and of humans of many races, theorized that between 140,000 and 290,000 years ago, fully modern human beings appeared in Africa in a more or less single event of speciation. Their appearance may have eventually displaced earlier human types, including the Neanderthals in Europe and evolved forms of an earlier human type known as *Homo erectus*.

The new humans, according to Stringer and Andrews, eventually expanded beyond Africa where, about 110,000 years ago, the Caucasoid (white) line split off from the African one. This split was followed, at about 40,000 years ago, by another split when the Mongoloids diverged from the Caucasoid mold, perhaps as a result of their isolation after colonizing the Far East.

Whatever the implications of these splits, Stringer and Andrews clearly felt they reflected the emergence of three races: the African-centered Negroid race, the European/West Asian–centered Caucasoid race, and the East Asian–centered Mongoloid race. Rushton embraced the theory, using it like the final piece of a jigsaw puzzle, one that brought his other findings into a satisfying evolutionary scheme. The three racial radiations clicked nicely into place with cranial capacity, IQ, sexual restraint, and all the rest.

To arrive at their out-of-Africa theory, Stringer and Andrews had worked backward from today's populations, comparing blood proteins and mitochondrial DNA. The strategy of tracing the past through the medium of present populations echoes Rushton's approach. But the Stringer and Andrews theory, if valid, only shows that today's three racial populations have inherited varying amounts of genetic material from an originally African population of humans without saying anything about what makes them different, either now or in the past. On one hand, their conclusions do not specifically prohibit quite different original "races" nor, on the other hand, does it explicitly exclude the possibility that the three "races" are pretty much as they were, from their origins.

The out-of-Africa theory of recent human evolution has had to slug it out in the anthropological literature with a competing theory called the multiregional model. In this theory, modern *Homo sapiens* emerged more or less everywhere from an ancestral population of an earlier human type called *Homo erectus*.

In citing the out-of-Africa theory for support, Rushton made a number of assumptions. First, he assumed that larger cranial capacity reflects an evolutionary advance. Second, he assumed that those characteristics such as cranial capacity, which have been used to gauge human evolution over millions of years, also apply in the span of tens of thousands of years. In support of this assumption, he assumed (third) that rather small changes (or differences) in cranial capacity had the same inferential validity as much larger ones.

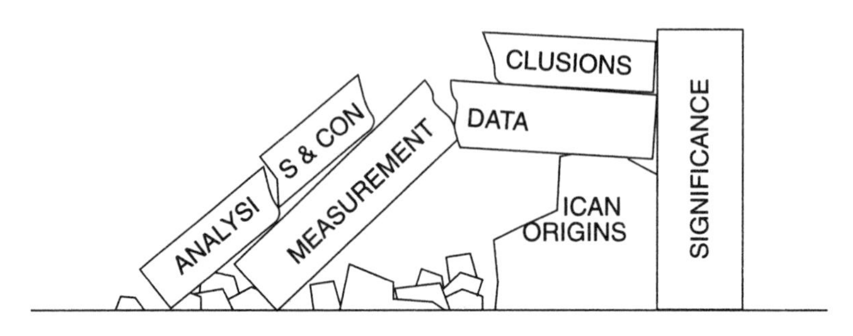

Rushton's theory.

As we are about to see, such small changes have no significance at all.

What Size Is Your Brain?

Rushton was hardly the first to study the brain sizes or IQs of different races. Instead, he is merely the latest in a line of researchers stretching back to the end of the nineteenth century. With the advent of Darwinian thought, overeager disciples of the theory sought to find confirmation of the struggle for fitness among the human races. They interpreted the theory with an astonishing naïveté, treating their technical civilization as the acme of human development and treating other, less "advanced" peoples as somehow tainted with primitivism.

Early enthusiasts of racial differences launched a campaign of measurement that seems to have been aimed at establishing European superiority over Africans, Aborigines, and all humans who enjoyed a more traditional lifestyle. This line of thought spawned not only the mistaken science of phrenology (a study in itself) but an orgy of measurement intended to map out the physical and mental characteristics of the races of humanity.

They measured everything that could be measured, from the length of each tarsal bone in the toes to the width, depth, and circumference of the skull. In the process they founded the practice called anthropometrics, a largely practical skill that is used today for a variety of legitimate purposes, from the study of human growth to the manufacture of clothing and ergonomic studies.

Prominent among the early anthropometricians was Sir Francis Galton, a distinguished British scientist, early statistician, and cousin of Sir Charles Darwin. In 1883 Galton coined the term "eugenics." He had a strong conviction that intelligence was heritable and that Britain should do everything in its power to encourage the production of genius. "Lesser breeds" received short shrift in Galton's intellectual mating game. Although he never defined what he meant by intelligence, Galton felt reasonably sure that it inhered to the greatest extent in the most widely recognized scientists and artists of his time. He studied families in which scientific or artistic brilliance seemed to prevail.

There can be no doubt that the size of a brain has some effect on its overall ability to function. Most neurophysiologists would not expect much from a human being with a brain the size of an earthworm, for example. And yet there have been humans of great accomplishment who had brains that were considerably smaller than the modern average.

The average European brain weighs, according to Gould, between 1,300 and 1,400 grams. And yet Walt Whitman, the famous American poet, had a brain of 1,282 grams while the French writer Anatole France had a brain that weighed a mere 1,017 grams, squarely within the range of *Homo erectus*, our ancient hominid ancestor!

One could go even further and twist the old adage a bit: Half a loaf is better than one. Consider the following case that is well known to all who study the human brain: In 1935 a female patient by the name of E.B. complained of severe headaches. Doctors discovered that her frontal lobes were badly calcified. A series of three operations resulted in the collective removal of almost the entire right half of E.B.'s brain. Nevertheless, not only did E.B. experience a more or less full recovery, she scored over 150 on the Stanford-Binet test (*after* the operation) and continued on to complete medical school!

In view of the plasticity of intelligence, it seems highly unlikely that such small differences in average cranial capacity could give rise to a systematic advantage. The famed Neanderthals who had cranial capacities fully the equal of modern Europeans are no longer with us. The (allegedly) small-brained Africans, on the other hand, continue to flourish.

The Bell Curves for You

Almost all measurements of humans, whether of body parts or performance and behavior, show the classic shape of the normal distribution also known as the bell curve. Hardly confined to human measurements, the normal distribution appears throughout nature. Surprisingly, mathematics itself may provide a clue why this is so. The Law of Large Numbers, a statistical theorem, states that any measurement with many contributing factors, each with its own statistical distribution, will tend to develop a normal distribution regardless of what the contributing distributions might be. Growth of bodies and body parts, being the cumulative outcome of many influences and factors, will therefore tend to express the combined influences, in all their variety, as a normal distribution.

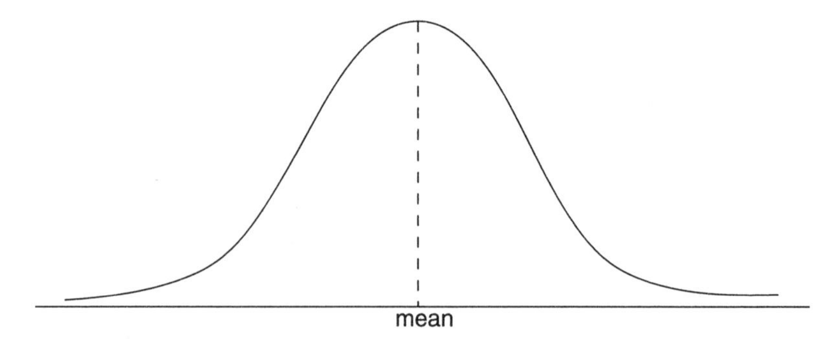

The bell curve.

The normal distribution's famous bell-shaped curve describes a population of numbers in which relatively many values cluster about the mean or average value and relatively few lie far away. From the mean at the center of the distribution to the two "tails" on the right and left, the number of values tapers gently away to nothing. As the word "distribution" implies, the height of the curve over any small segment of the horizontal axis indicates how many individuals' measurements are distributed within that segment. To find the proportion of such individuals, simply multiply the width of the segment by the height of the curve above it. A small segment in the center of the curve obviously has more individual values piled within it than an equal-sized segment that is far out in one of the tails.

In practice, a large number of measurements of a single variable, whether of brain size or any other measurable quantity, does not yield a smooth, continuous curve like the one shown in the figure, but the choppy, discrete shape of a histogram. The idea is very similar, however, because within each small range of values that appears along the bottom of the histogram, the number of individual observations that fall within that range are stacked up like so many coins.

The histogram in the figure on the next page, for example, displays the distribution of heights of a large sample of German males. Along the horizontal axis, the height variable ranges from a short 150 centimeters (about 4 feet 11 inches) to a tall 200 centimeters (about 6 feet 7 inches). Each of the fifty subranges spans one centimeter of height and the vertical bar above it represents the proportion of individuals whose height lies within that one centimeter subrange. If you look at the bar centered on the 161 subrange (just to the right of the one labeled 160), for example, you may read off the percentage of German men in the sample whose height was measured to be anywhere from 161.0 to 161.9 centimeters, approximately one percent in this case. In other words, about one percent of the men sampled had heights in this range.

The shape of the histogram in the figure is only approximately normal, a feature shared by all measurements of normally

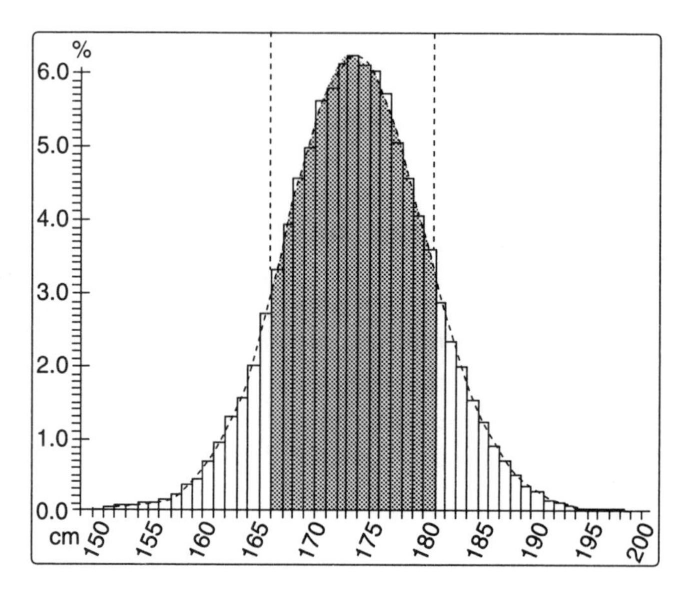

The distribution of heights for German males.
(Adapted from *Der "Vermessene" Mensch.*)

distributed populations. You can expect a histogram of real-world data almost never to match perfectly a theoretical underlying distribution, just as you can expect that throwing one hundred pennies in the air will almost never result in exactly fifty heads and fifty tails, the underlying theoretical distribution.

When a scientist wishes to compare two populations, he or she must first gather samples as surrogates for the two populations, then apply statistical tests that measure the extent of overlap, in effect.

In the next figure I have illustrated this overlap by using two pairs of normal curves. These represent two populations of cranial capacities, the one on the left for blacks, the one on the right for whites. These curves do not come from Rushton's papers, but represent my own reconstruction of the entire black and white populations of the United States. There are so many people altogether, that the histograms are indistinguishable at this resolution, from continuous curves. I have used the bell curve as a reasonable approximation to the real populations.

All of Rushton's samples are gleaned from the anthropometric literature, including the sample on which I have based

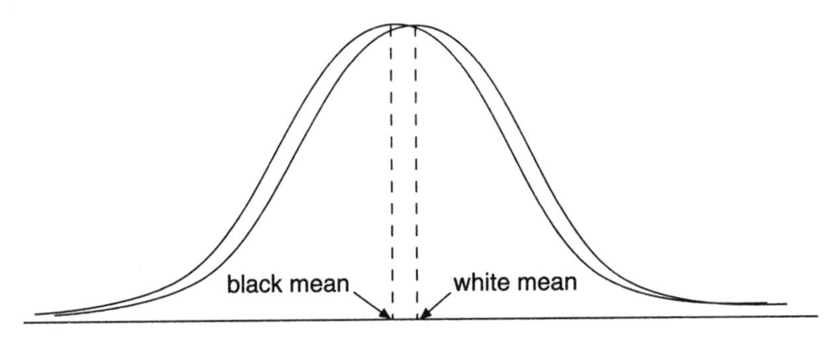

Black and white cranial capacities.

these curves. In 1988 the U.S. Army commissioned a survey of head sizes of its personnel, presumably in order to know what sizes of helmets, caps, and hats to order. Using a formula first developed in 1901, Rushton converted all of the head measurements in the army survey (length, height, and breadth of skull) into cranial capacities. He then carried out an extensive series of statistical analyses on the resulting numbers. Among the "significant" differences he uncovered were mean cranial capacities of 1,378 cubic centimeters for the 2,871 Caucasoids surveyed and 1,362 cubic centimeters for the 2,676 Negroids.

I have used these figures for the averages that appear in the figure. The difference in these means, a mere 16 cubic centimeters, is reflected in the figure by the minute shift that separates the two normal curves. Let us suppose for the moment that these curves reflect the actual distribution of cranial capacities among Caucasoids and Negroids.

A close examination of the two cranial capacity distributions enables us to draw some interesting conclusions, not about the differences but about the similarities of both black and white populations. First, the immense overlap ensures that in something like 95 percent of both populations, the cranial capacities are exactly the same. For each individual in one population you will be able to find an individual in the other population with an identical cranial capacity.

In fact, because the two sample sizes are nearly identical, you could arrange for each person in one sample to have a "cranial buddy" in the other. Since the two distributions do not overlap completely, there would have to be a little doubling up on both sides of the distributions. On the low side, for example, a small percentage of whites might have to accept two cranial buddies instead of one. The same thing would be true of blacks on the high side.

This simple fact of distributions illustrates clearly what "within-group" and "between-group" differences are all about in connection with the brain size controversy raised by Rushton's work. In particular, it shows that as far as these two populations are concerned, within-group differences simply swamp between-group differences. For this reason alone, any public policy based on such a strongly overlapping difference would obviously be unfair. For example, if it was decided that any black person with a cranial capacity of less than so many cubic centimeters should be given some kind of special treatment (or nontreatment), then the very same treatment would have to be given to the many whites with the same cranial capacity.

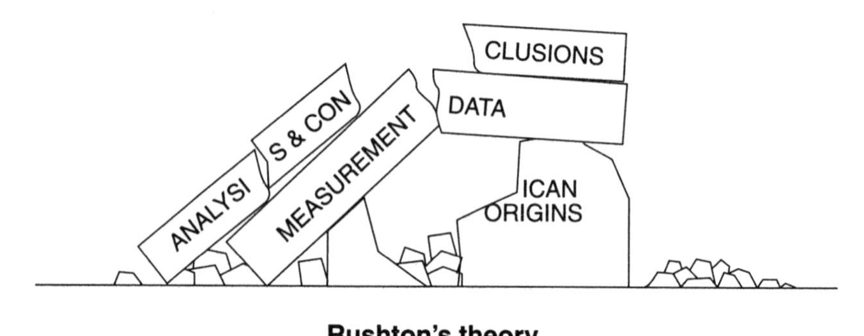

Rushton's theory.

Cleaning the Castle

J. Phillipe Rushton was not the first to develop a theory of racial differences, nor is he the last. The publication in 1994 of *The Bell Curve: Intelligence and Class Structure in American Life* by psychologist Richard Herrnstein and social scientist Charles Murray reopened the issue with the same, by now predictable, results. The book repeats many of the same studies undertaken by Jensen and Rushton and reaches many of the same conclusions. At the same time, the media produced many of the same, superficial and off-center analyses. As a glance at *The Bell Curve* reveals, the apprentices are at it again, incanting page after page of magic formulas in the hope that Truth herself will appear.

It is truly time that psychology finds some new brooms to sweep clean its corner of the old castle.

Further Reading

The following books and articles provide valuable background reading for each of the cases of bad science explored in this book. Some of these references argue that the science is bad and some argue the opposite, at least in cases that remain controversial.

1. The Century Begins

William Seabrook. *Dr. Wood, Modern Wizard of the Laboratory.* New York: Harcourt Brace, 1944.

Contains an account of Wood's visit to Blondlot's laboratory.

Mary Jo Nye. "N-rays: An Episode in the History and Psychology of Science." In *Historical Studies in the Physical Sciences,* Vol. 11, Number 1, 1980, pp. 125–156. The University of California, Berkeley.

Perhaps the most detailed account of the N-ray debacle published to date.

Irving Langmuir. "Pathological Science." (trans. and ed. Robert N. Hall). In *Physics Today,* October, 1989, pp. 36–44.

A review of Langmuir's laws of pathological science.

2. Mind Numbers

David Wechsler. *The Range of Human Capacities.* New York: Hafner Publishing, 1969.

Wechsler defends intelligence as a scale of innate human capacity.

Carl C. Liungman. *What Is IQ? Intelligence, Heredity and Environment.* London: Gordon Cremonesi Ltd., 1970.

Among other things, this book describes all the major tests.

N. J. Block and Gerald Dworkin (eds.). *The IQ Controversy—Critical Readings.* New York: Pantheon Books/Random House, 1976.

A collection of readings from a wide variety of contributors.

Paul L. Houts (ed.). *The Myth of Measurability.* New York: Hart Publishing, 1977.

The best collection of readings critical of the IQ concept.

R. C. Lewontin, Steven Rose, and Leon J. Kamin. *Not In Our Genes: Biology, Ideology and Human Nature.* New York: Pantheon Books, 1984.

Lewontin and company explain genes and their expression in the human body and brain.

Stephen Jay Gould. *The Mismeasure of Man.* New York: W. W. Horton & Company, 1993.

A detailed but engaging account of the history of attempts to measure human mental capacity, from craniometry to IQ.

3. Dreaming Up Theories

Adolf Grünbaum. *The Foundations of Psychoanalysis: A Philosophical Critique.* Berkeley/Los Angeles: University of California Press, 1984.

Required reading for those who may have wondered about the dark side of psychoanalytic theory and its origins.

Susan Allport. *Explorers of the Black Box.* New York: W. W. Norton & Company, 1986.

An account of the difficulties and triumphs encountered by neurophysiologists trying to understand the simplest brains.

Frank J. Sulloway. "Reassessing Freud's Case Histories." In *ISIS,* Vol. 82, Number 312, June 1991, pp. 245–275.

A summary of Sulloway's analysis of Freud's six published case studies.

Frank J. Sulloway. *Freud, Biologist of the Mind.* Cambridge, Mass.: Harvard University Press, 1992.

Completed during Sulloway's conversion to the Grunbaum perspective on Freud, this book is nevertheless crammed with fascinating detail about Freud and his work.

4. Surfing the Cosmos

Philip Morrison, John Billingham, and John Wolfe (eds.). *The Search for Extraterrestrial Intelligence.* Washington, D.C.: National Aeronautics and Space Administration, Ames Research Center, 1977.

A well-written account of the speculations and technical details that have informed the SETI project.

David W. Swift. *SETI Pioneers.* Tucson: University of Arizona Press, 1990.

A fascinating look at the personal motivations of many SETI "pioneers."

Ben Bova and Byron Preiss (eds.). *First Contact: The Search for Extraterrestrial Intelligence.* New York: NAL Books/Penguin, 1990.

A book full of science fiction and other whacky perspectives.

Frank Drake and Dave Sobel. *Is Anyone Out There? The Search for Extraterrestrial Intelligence.* New York: Delacorte Press, 1992.

Drake's heartwarming account of his yearning for extraterrestrial companionship.

Paul Horowitz and Carl Sagan. "Five Years of Project META: An All-Sky Narrow-Band Search for Extraterrestrial Signals." In *The Astrophysical Journal,* Vol. 415, Sept. 1993, pp. 218–235.

5. The Apprentice Builds a Brain

Frank Rosenblatt. *Principles of Neurodynamics: Perceptions and the Theory of Brain Mechanisms.* Washington, D.C.: Spartan Books, 1961.

Rosenblatt's attempt to define a new science, or at least, a technology.

Marvin Minsky and Seymour Papert. *Perceptrons.* Cambridge, Mass.: MIT Press, 1969.

The book that nearly killed neural net research.

David E. Rummelhart and James I. MacLelland (eds.). *Parallel Distributed Processing.* Cambridge, Mass.: MIT Press, 1986.

A rambling collection of mostly gung-ho articles about neural net approaches.

Francis Crick. "The Recent Excitement About Neural Networks." In *Nature,* Vol. 337, 1989, pp. 129–132.

In this article Crick seems to draw back from his earlier enthusiasm for neural nets.

6. Genie in a Jar

Eugene F. Mallove. *Fire From Ice: Searching for the Truth Behind the Cold Fusion Furore.* New York: John Wiley & Sons, 1991.

The best exploration of cold fusion from the believer's side.

John R. Huizenga. *Cold Fusion: The Scientific Fiasco of the Century.* Rochester, N.Y.: University of Rochester Press, 1992.

The best account of behind-the-scenes details of the cold-fusion story by a key participant.

Gary Taubes. *Bad Science: The Short Life and Weird Times of Cold Fusion.* New York: Random House, 1993.

A standard reference for the bad science behind cold fusion.

7. Biosphere 2 Springs a Leak

John Allen. *Biosphere 2: The Human Experiment.* New York: Viking Penguin, 1991.

You could have bought this coffee-table treat at the Biosphere 2 gift shop on opening day.

Johnny Dolphin. *The Dream and Drink of Freedom.* Bonsall, Calif.: Synergetic Press, 1987.

The poem "Lebensraum" quoted on page 136 may be found in this work.

Dorion Sagan and Lynn Margulis. *Biospheres: From Earth to Space.* Hillside, N.J.: Enslow Publishers, 1989.

A wide-eyed look at the possibilities for human space colonies with a special preview of Biosphere 2.

8. For Whom the Bell Curves

Richard Leakey and Roger Lewin. *Origins Reconsidered: In Search of What Makes Us Human.* New York: Doubleday, 1992.

A fascinating glimpse of human prehistory by two leading experts.

Alan Bilsborough. *Human Evolution.* Tertiary Level Biology Series. London: Blackie Academic & Professional, 1992.

More details for fans of early hominids.

Stephen Jay Gould. *The Mismeasure of Man.* New York: W. W. Norton & Company, 1993.

A detailed but engaging account of the history of attempts to measure human mental capacity, from craniometry to IQ.

Richard J. Herrnstein and Charles A. Murray. *The Bell Curve: Intelligence and Class Structure in American Life.* New York: The Free Press, 1994.

This classic of cargo-cult science will handsomely repay any reader fascinated by pages of formulas.

J. Phillipe Rushton. *Race, Evolution, and Behavior: A Life History Perspective.* New Brunswick, N.J.: Transaction Publishers, 1995.

Rushton's culminating account of racial differences as revealed by a great many statistical analyses.

Acknowledgments

An able research assistant lends wings to the nonfiction keyboard. I thank Anna-Lee Pittman for assistance with the early searches on Biosphere 2 and cold fusion, as well as on Freud, psychiatry, and IQ. I thank Patricia Dewdney for many useful discussions on reference strategies. I thank Victor Dricks of the *Phoenix Gazette*, Marc Cooper of the *Village Voice*, Frieda B. Taub of the University of Washington in Seattle, and Abdelhaq Hamza of the University of New Brunswick, for explorations of the scientific and human issues.

Index